SKY-
SCRAPER

摩天大楼

赵波 王翠萍 译

高迪国际 HI-DESIGN PUBLISHING 编

大连理工大学出版社

图书在版编目 (CIP) 数据

摩天大楼：英汉对照 / 高迪国际 HI-DESIGN
PUBLISHING 编；赵波，王翠萍译. —大连：大连理工
大学出版社，2013.10
　　ISBN 978-7-5611-8225-3

　　Ⅰ.①摩… Ⅱ.①高… ②赵… ③王… Ⅲ.①高层建
筑－建筑设计－英、汉 Ⅳ.① TU972

　　中国版本图书馆 CIP 数据核字 (2013) 第 216726 号

出版发行：大连理工大学出版社
　　　　　（地址：大连市软件园路 80 号 邮编：116023 ）
印　　刷：上海锦良印刷厂
幅面尺寸：240mm×320mm
印　　张：21
插　　页：4
出版时间：2013 年 10 月第 1 版
印刷时间：2013 年 10 月第 1 次印刷
策划编辑：袁　斌　刘　蓉
责任编辑：刘　蓉
责任校对：李　雪
封面设计：张　吟

ISBN 978-7-5611-8225-3
定　　价：338.00 元

电话：0411-84708842
传真：0411-84701466
邮购：0411-84703636
E-mail:designbookdutp@gmail.com
URL:http://www.dutp.cn

如有质量问题请联系出版中心：（0411）84709246 84709043

Carlos Rubio Carvajal

The skyscraper is the combination of all the technologies of the past 150 years, providing a first class architectural phenomenon that shapes the skyline of the growing cities. Its construction still causes big interest and fascination in society, concerning and involving landscapers, architects, critics, urban planners, politicians, sociologists, economists and ecologists.

The reason why big companies have historically tended to cluster in these kind of buildings is because the concentration of people and resources in small areas allows more efficiency and productiveness, saving costs of internal and external messaging, as skyscrapers are grouped at the center of major cities, allowing easy connection to the public transportation for employees, messengers and clients.

But changes in information technology in the last two decades and the World Wide Web, along with the increasing mobility of employees and the proliferation of smartphones and tablets have changed the concept of business, transforming it into a global company, arising new organizational schemes. Employees have achieved greater independence of space and time which have facilitated, in some cases, big firms dispense headquarters in the City Downtown.

For a long time, the working world has been characterized by rigid schedules, fixed sites and central structures. Now the relaxation of these parameters, lets us decide "who, where and when we work" allowing multiple customizable advantages, helping encouraging individual creativity, connectivity, competitiveness and performance, saving financial costs, time, even in ecological terms.

The aforementioned advantage of a profitable vertical grouping which concentrates resources, cutting costs and efforts, is also transferable to a bigger living organism as "The City". A thermodynamic analysis of the city supports the capital importance of density versus the dispersed city that spreads like an oil slick. "The compact city" is today the preferred model for landscape architects, economists and sociologists. It is the most sustainable model and the better able to attract talent, generate innovation and competence. It is also the most effective tool for addressing the current economic crisis.

Carlos Rubio Carvajal
Co-Lead Partner of Rubio & Álvarez-Sala Architecture

摩天大楼集合了过去150年来的所有的技术，展现了一个一流的建筑体量，塑造了日益发展的城市天际线。摩天大楼的建设还在继续引起社会的极大兴趣和关注，尤其是对参与其中的景观设计师、建筑师、评论家、城市规划者、政治家、社会学家、经济学家和生态学家们而言。

从历史上看，大型企业都倾向于聚集在这样的建筑中。因为在小范围内集中人力和各种资源，能提高效率和生产力，节省内外部信息传达的成本。而且摩天大楼分布在大城市的中心，可以方便公司员工、邮递员和客户乘坐公共交通工具。

但是，随着过去二十年来信息技术和互联网的发展变化，员工流动性的增加，以及智能手机和平板电脑的普及，商业的概念已经改变，出现了全球化的公司和新的组织结构。雇员的办公空间和时间更灵活，在有些情况下，大型企业可以不把总部设在市中心。

很长一段时间里，职场的特点表现为严格的时间表、固定的场所和中央组织结构。现在这些要求正在放宽，在决定"和谁、何地、何时工作"时，允许员工发挥更多可变通的优势，帮助鼓励每个人去提高创造力、协同性、竞争力和绩效，节约在财务、时间，甚至是生态方面的成本。

上文提到的摩天大楼拥有的各种优势，如集中了资源，减少了成本和付出，也可以转移到"城市"这个更大的生物体上来。城市的热力学分析支持资本密度高的城市，而不是像水面浮油一样扩散的分散型城市。今天，"紧凑型城市"是景观建筑师、经济学家和社会学家的首选模式。它是最可持续发展的模式，能够更好地吸引人才，激发创新能力，也是解决当前经济危机最有效的工具。

卡洛斯·卢比奥·卡瓦哈尔
Rubio & Álvarez-Sala 建筑事务所 联席合伙人

CONTENTS

目 录

DUKE ENERGY CENTER

杜克能源中心

LOCATION:
NORTH CAROLINA, USA

ARCHITECT: TVSDESIGN
CLIENT: WACHOVIA BANK
AREA: 139, 354 m²

Serving as a new gateway to downtown, the project provides a vibrant new urban focus for the city. As with all great cities around the world, the complex will provide an enriching, diverse experience for the citizens of Charlotte and its visitors. With opportunities for work, shopping, and a multiplicity of diverse cultural offerings, the project lifts the city to new heights.

The initiatives that Wachovia has incorporated involve a whole building approach to address areas of human and environmental health including sustainable site development, increased water and energy savings, responsible use of materials and resources, as well as, providing a premium, quality indoor environment.

The crystalline form of the tower produces a distinctive modern image reflective of this dynamic growing city. An eight level underground parking deck connects to three adjacent city blocks and provides a base for two of the museums. tvsdesign provided master planning, overall project coordination.

1 MECHANICAL
2 OVERRIDE
3 OFFICE
4 PENTHOUSE
5 MECHANICAL
6 WAKE FOREST
7 SERVER LEVEL
8 CHURCH STREET

9 RETAIL
10 PARKING ENTRY
11 SERVER LEVEL
12 LOBBY
13 TRYON STREET
14 OFFICE SERVICE DOCK

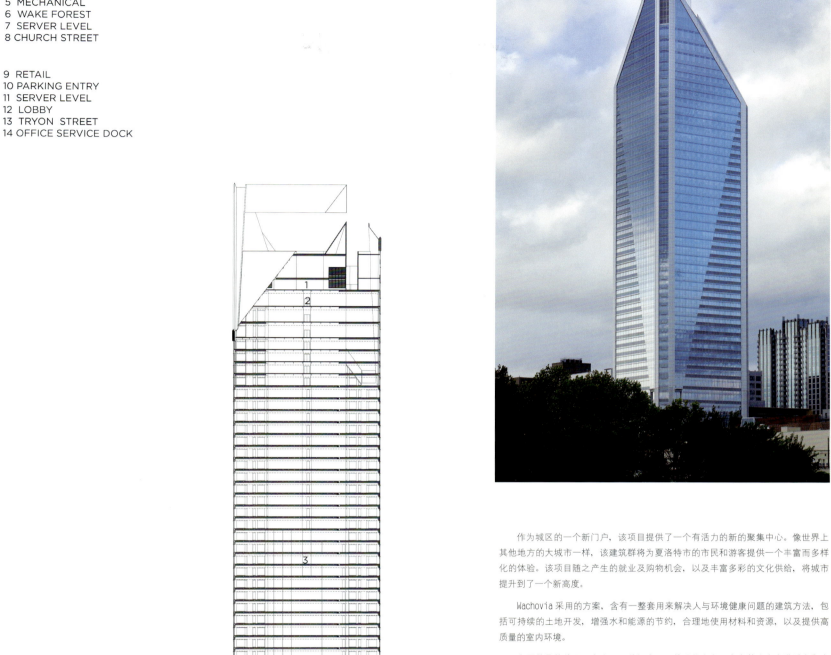

作为城区的一个新门户，该项目提供了一个有活力的新的聚集中心。像世界上其他地方的大城市一样，该建筑群将为夏洛特市的市民和游客提供一个丰富而多样化的体验。该项目随之产生的就业及购物机会，以及丰富多彩的文化供给，将城市提升到了一个新高度。

Wachovia 采用的方案，含有一整套用来解决人与环境健康问题的建筑方法，包括可持续的土地开发，增强水和能源的节约，合理地使用材料和资源，以及提供高质量的室内环境。

大厦的晶体外观，产生了一种与众不同的现代印象，寓意着这座充满活力和发展的城市。大厦地下是一个八层的停车场，停车场与三个相邻的街区连接，同时也是两个博物馆的停车场。 tvsdesign 负责了总体规划和项目的总体协调。

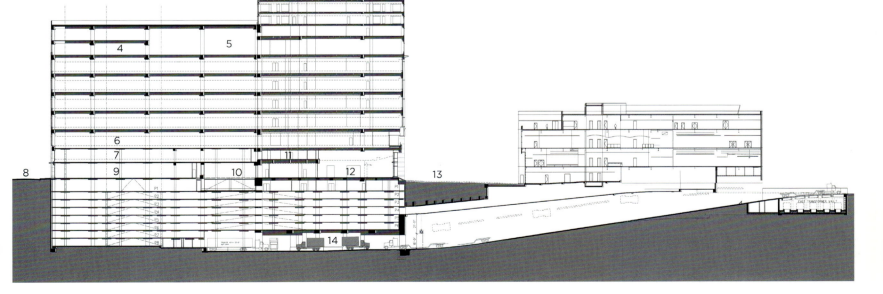

0 20 50 100 ft

SECTION **A**
LOOKING NORTH THROUGH SERVICE DRIVE

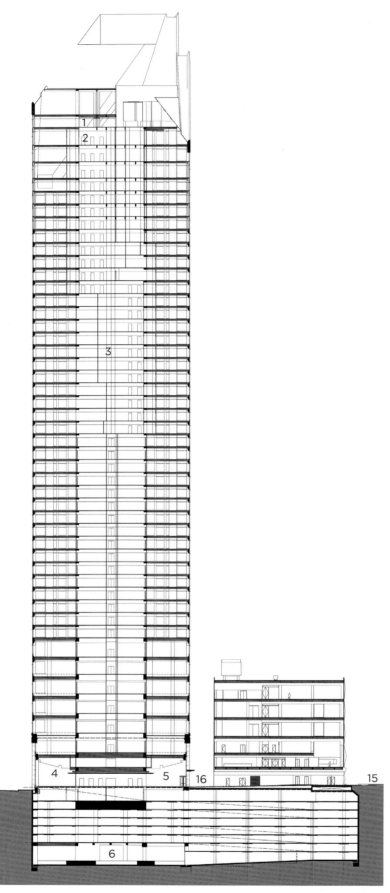

SECTION **B**
LOOKING NORTH THROUGH SERVICE DRIVE

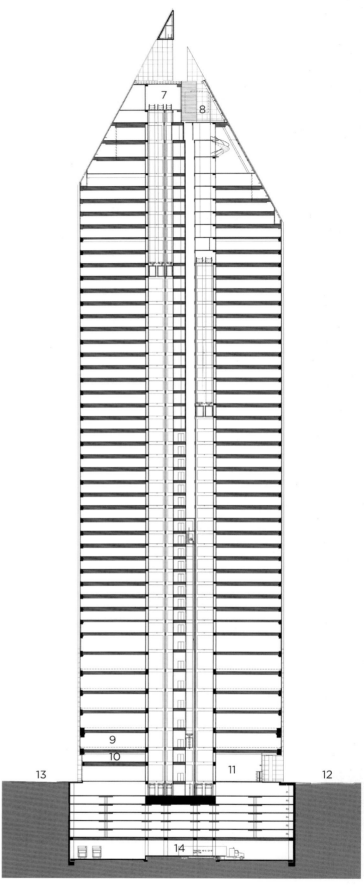

SECTION **C**
LOOKING NORTH

0 20 50 100 ft

1 MECHANICAL	9 WAKE FOREST
2 OVERRIDE	10 SERVER LEVEL
3 OFFICE	11 THEATER
4 LOBBY	12 TRYON STREET
5 RETAIL	13 CHURCH STREET
6 SERVICE LEVEL	14 MID-RISE CORPORATE OFFICE
7 ELEVATOR MACHINE	15 FIRST STREET
8 ROOF WELL	16 RETAIL STREET

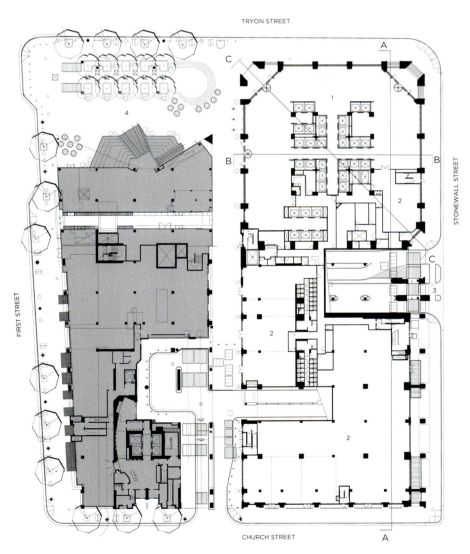

TRYON STREET

CHURCH STREET

FIRST STREET

STONEWALL STREET

A

B

C

C

D

1

2

2

2

4

3

1 CORPORATE OFFICE LOBBY
2 TENANT RETAIL
3 STONEWALL PARKING ENTRANCE
4 OUTDOOR PLAZA/ HARDSCAPE

NOT IN SCOPE

LEVEL 1 FLOOR PLAN

NORTH   0   15   30   60 ft

MODE GAKUEN COCOON TOWER

日本东京蚕茧大厦

LOCATION: TOKYO, JAPAN

ARCHITECT: TANGE ASSOCIATES
STRUCTURAL ENGINEER: ARUP JAPAN
MEP ENGINEER: KENCHIKU SETSUBI SEKKEI KENKYUSHO

CONTRACTOR: SHIMIZU CORPORATION
CLIENT: MODE GAKUEN
AREA: 80,865.40 m²

PHOTOGRAPHER: KOJI HORIUCHI

Mode Gakuen Cocoon Tower, which contains 3 vocational schools with approximately 10,000 students, is an innovative educational facility located in Tokyo's distinctive Nishi-Shinjuku high-rise district. The building's elliptic shape, wrapped in a criss-cross web of diagonal lines, embodies the "cocoon" concept. Students are inspired to create, grow and transform while embraced within this cocoon-like, incubating form.

In essence, the creative design successfully nurtures students to communicate and think creatively. In designing Mode Gakuen Cocoon Tower, Tange Associates offers a new solution for school architecture in Tokyo's tightly meshed urban environment. A new typology for educational architecture, the tower and accompanying auditoriums successfully encompass environmental concerns and community needs with an unparalleled inspirational design.

The high-rise tower floor plan is simple. Three rectangular classroom areas rotate 120 degrees around the inner core. From the 1st to the 50th floor, these rectangular classroom areas are arranged in a curvilinear form. The inner core consists of an elevator, staircase and shaft. Unlike the typical horizontally laid out school campus, the limited size of the site challenged Tange Associates to develop a new typology for educational architecture.

Student Lounges are located between the classrooms, facing three directions, east, southwest and northwest. Each atrium lounge is three-storey high and offers sweeping views of the surrounding cityscape.

A new types of schoolyard, these innovative lounges offer students a comfortable place to relax and communicate. The elliptic shape permits more ground space to be dedicated to landscaping at the building's narrow base, while the narrow top portion of the tower allows unobstructed views of the sky.

The nurturing forces of nature are close at hand to the student, an inspiring environment in which to study, learn and grow. For the community, the fascinating design of Mode Gakuen Cocoon Tower is a welcome contribution to the urban landscape and an example of how such design innovation benefits and impacts its immediate surroundings.

蚕茧大厦位于东京著名的西新宿区的超高层建筑群区，是一个创新型的教学建筑，内有三个职业学校，容纳了约10000名学生。建筑的椭圆形外形包裹在层层相互交叉的网状结构中，体现了"蚕茧"这一主题概念。学生在这个蚕茧一样的"孵化器"中学习、成长，创新能力也得到启蒙。

实际上，这种创新设计成功地培养了学生的交际能力和创新思维。在大厦的设计过程中，Tange Associates建筑公司在东京密集的建筑环境中挖掘了学校建筑的新空间。蚕茧大厦是一种新型的教育建筑，它和附属的礼堂成功考虑了周边环境和社区的需求，为我们呈现了无与伦比的灵感设计。

这座超高层建筑的平面图其实还算简单。三个矩形的教学区绕着大楼的中心，相互的角度差为120度。从一层到50层，这些矩形教室按照曲线分布。大楼的中心设有电梯、楼梯和竖井。由于不是传统的水平布局的校舍，Tange Associates建筑公司必须打破空间尺寸的限制，设计出新型的教学设施。

学生休息区位于教室和教室之间的空隙地带，分别位于大厦的东部、西南部和西北部。中庭的每个休息区都有三层楼高，可以鸟瞰周边城市风貌的全景。

这种创新的休息区是一种新型的校园活动场所，为学生提供了放松与交际的舒适场所。椭圆形的空间设计也给大楼的窄基底提供了更多的景观空间，而较窄的圆形楼宇顶部也让视野更开阔，天空的景色一览无余。

对学生来说，这是一个学习、研究、生活的灵性之地，能近在咫尺地感受到大自然的滋养。对于社区而言，蚕茧大厦的设计令人陶醉，为城市景观增添了亮色，也成为创新设计的典范，为周边环境带来了巨大的积极影响。

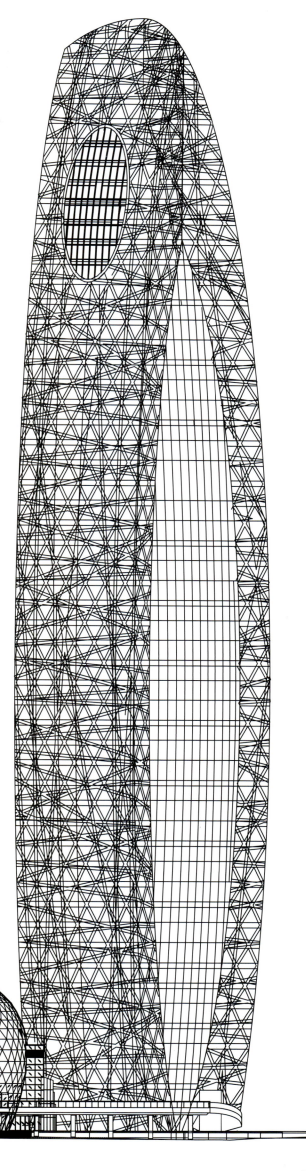

NORTH ELEVATION

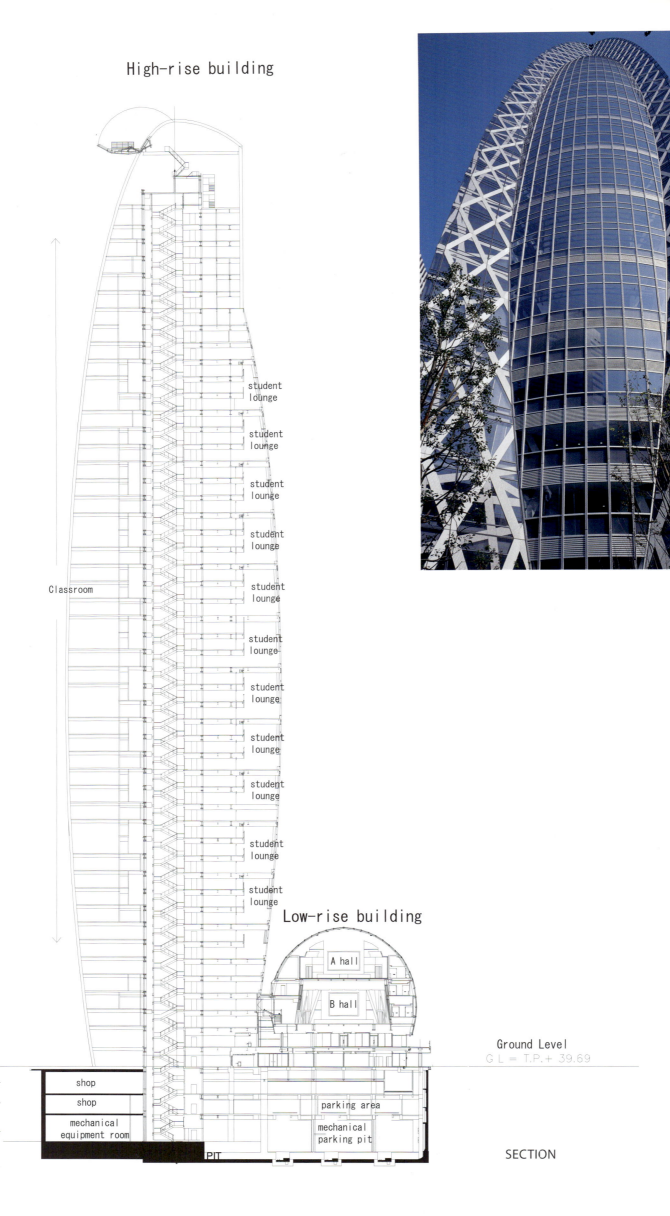

High-rise building

Classroom

student
lounge

student
lounge

student
lounge

student
lounge

student
lounge

student
lounge

student
lounge

student
lounge

student
lounge

student
lounge

student
lounge

Low-rise building

A hall

B hall

Ground Level
G L = T.P. + 39.69

B2F   shop

B3F   shop

B4F   mechanical
       equipment room

PIT

parking area

mechanical
parking pit

SECTION

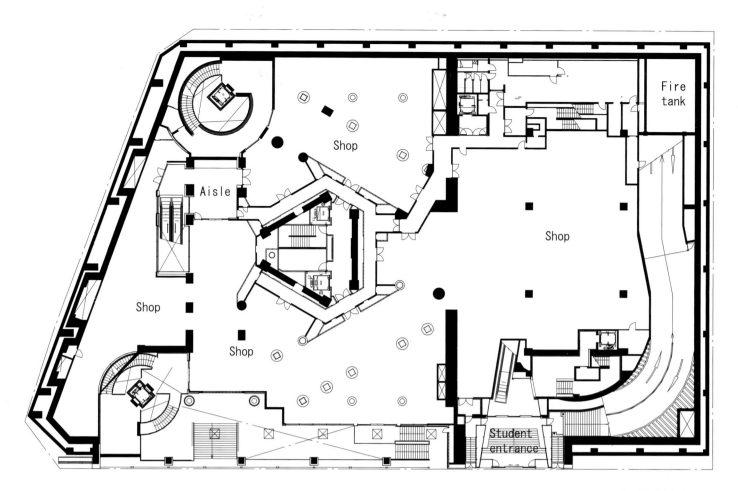

B2 FLOOR PLAN

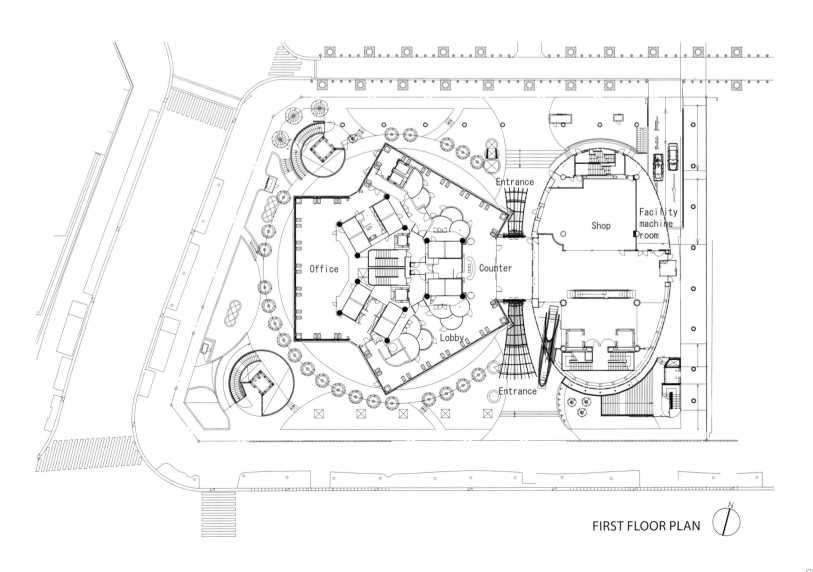

FIRST FLOOR PLAN

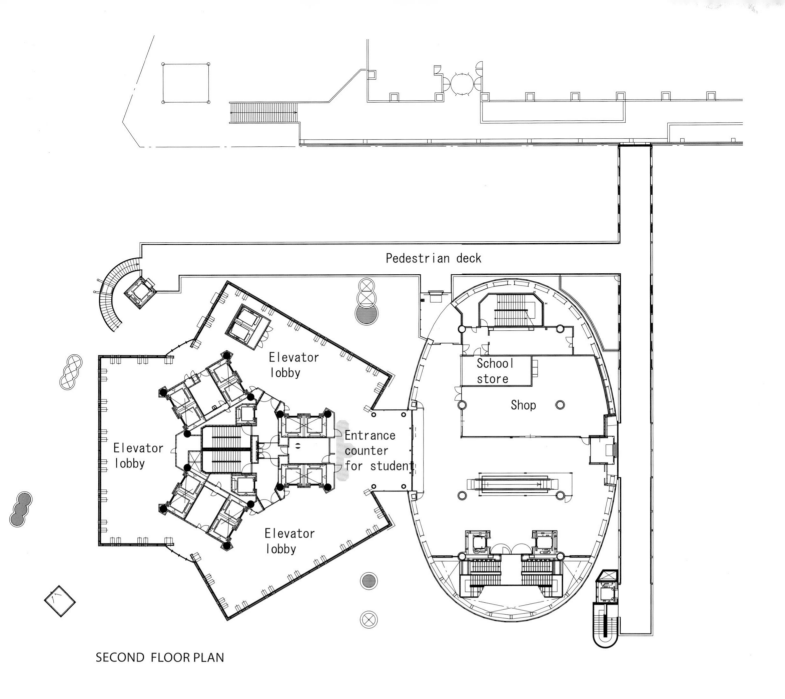

Pedestrian deck

Elevator lobby

Elevator lobby

Elevator lobby

School store

Shop

Entrance counter for student

SECOND FLOOR PLAN

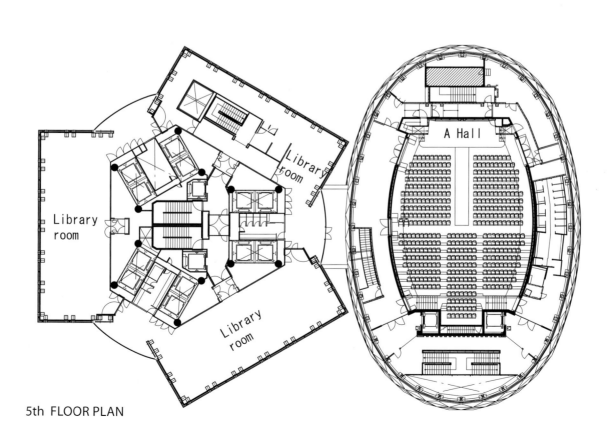

Library room

Library room

Library room

Library room

A Hall

5th FLOOR PLAN

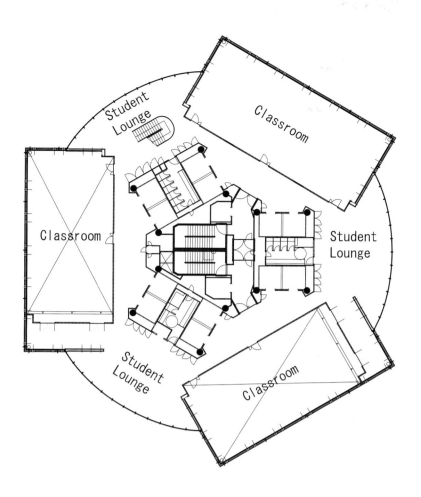

21st FLOOR PLAN

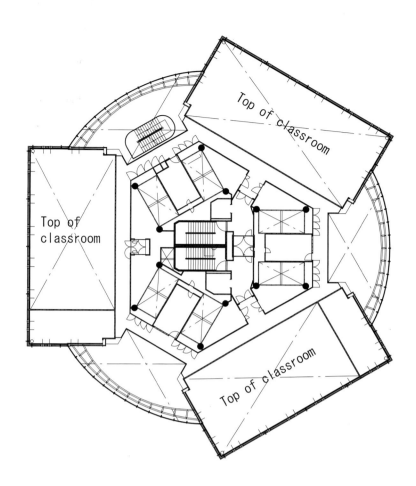

22nd FLOOR PLAN

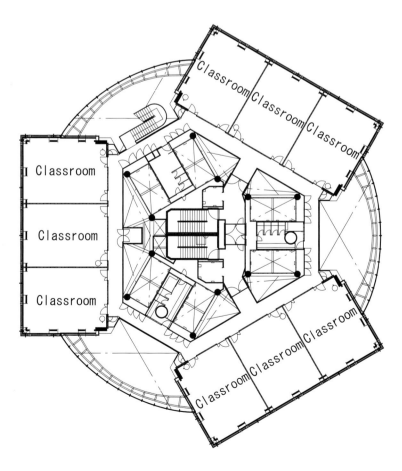

23rd FLOOR PLAN

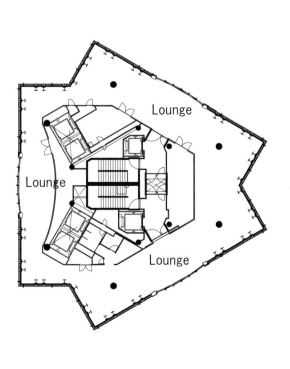

50th FLOOR PLAN

THE AL BIDDA TOWER

Al Bidda 大厦

LOCATION:
DOHA, QATAR

ARCHITECT: GHD
AREA: 5,552 m²
PHOTOGRAPHER: BRENDAN TEXEIRA, ALEX ATIENZA

The Al Bidda Tower dominates the skyline of Doha's central business area and Corniche. Its form blazes out of the desert sands in a striking, continually twisting face that depicts clear form in structure and elevation. Architecturally achieved by rotating the form at every level the design represents a continually evolving culture and economy. Functioning as a commercial office tower, tenants occupy an environment of unsurpassed modernism.

The Tower is divided into 3 sections. The Ground floor of area 1,860m² accommodates a double height entrance lounge and has provision for the administration offices of the tower and possible utilisation as meeting rooms, showrooms, art galleries, a business center and other usage as required and

identified by the client. It also houses a restaurant with access to an outdoor café in a landscaped area adjacent to the water feature adjoining the Tower. The Mezzanine floor provides further space for possible meeting rooms, multipurpose hall, business center, and cafeterias for the exclusive use of tenants. Level 1 ~ 39 inclusive provide for rentable office spaces.

The unique form of the building called for a structural system that would accommodate the variable floor plates as well as the shift of the 1.5 degrees of the apex of the triangular rotor shaped plan. The result is an avant-garde diagonal system that mimics the structural triangular truss work of the facade.

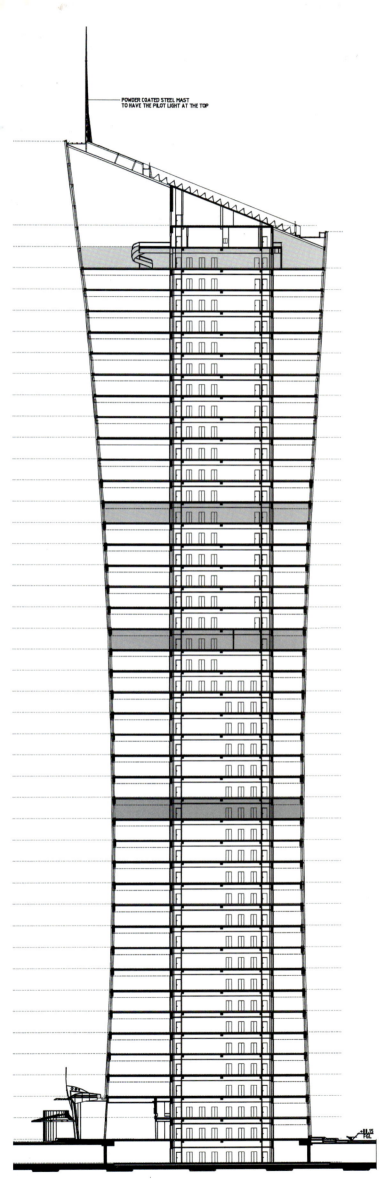

POWDER COATED STEEL MAST
TO HAVE THE PILOT LIGHT AT THE TOP

SECTION

Al Bidda 大厦位于多哈中心商务区，沿着海滨大道，耸立于天际。在沙漠强光的照射下，它的令人震撼的、不断向上盘绕的外观，让人可以清晰地看到它的结构和外观。大厦外形的每个角度都看似旋转，这样的建筑设计代表着不断发展的文化和经济。作为一栋商业办公楼，它的住户拥有了无与伦比的现代环境。

大厦共分为三部分。底楼面积1860平方米，涵盖了一个两层楼高的休息厅以及大楼行政管理办公室、会议室、陈列室、艺术画廊、商业中心等满足客户需求的各种空间。此外，底楼还设有一间餐厅，餐厅有通道与楼外风景区内的咖啡馆相连，且毗邻大楼的水景区。夹层楼提供了更多用作会议室、多功能厅、商务中心及自助餐厅的空间，以满足租户的特殊需求。1 层到 39 层为可租用的办公空间。

大厦独特的外形设计要求一个相对应的结构体系，该体系需要能够适应逐渐变化的楼面，以及三角旋转形顶点的 1.5 度的移位，最终形成了一个前卫的对角线系统，而该系统则模仿了三角形桁架结构的工艺。

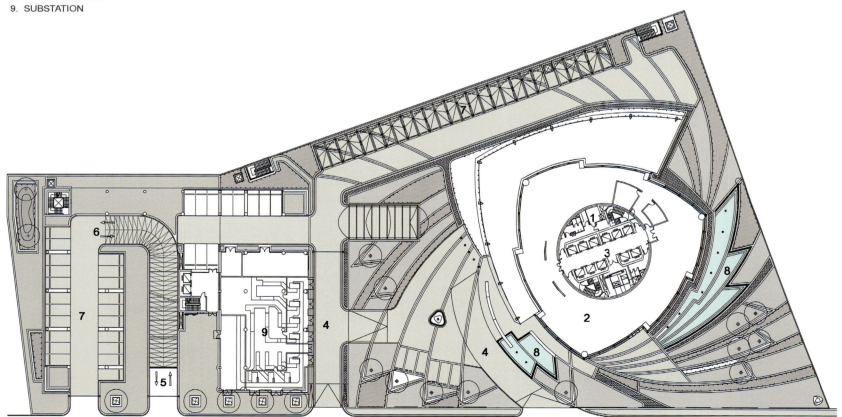

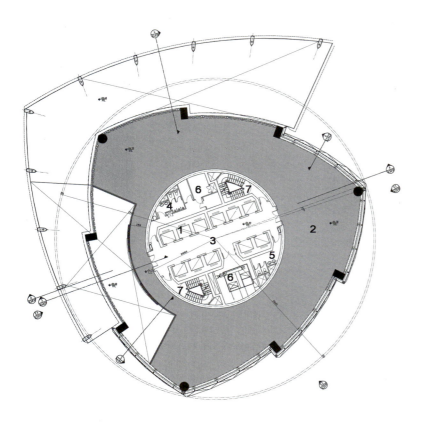

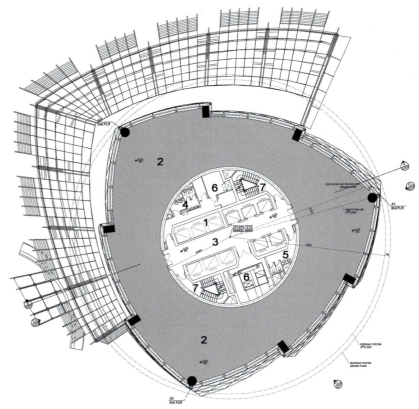

MEZZANINE LEVEL PLAN

LEGEND
1. CORE
2. FLOOR PLATE
3. ELEVATOR LOBBY
4. MALE TOILETS
5. FEMALE TOILETS
6. SERVICES
7. FIRE STAIRS

SECOND LEVEL PLAN

LEGEND
1. CORE
2. FLOOR PLATE
3. ELEVATOR LOBBY
4. MALE TOILETS
5. FEMALE TOILETS
6. SERVICES
7. FIRE STAIRS

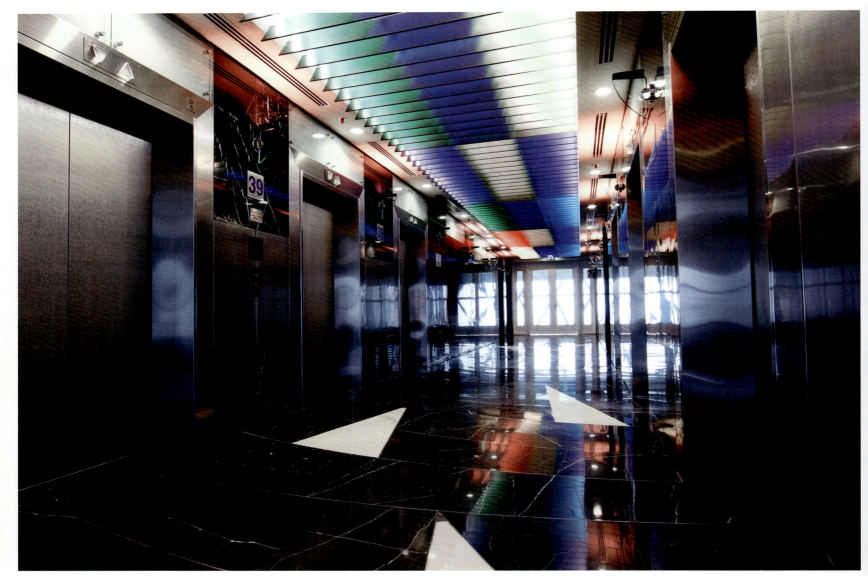

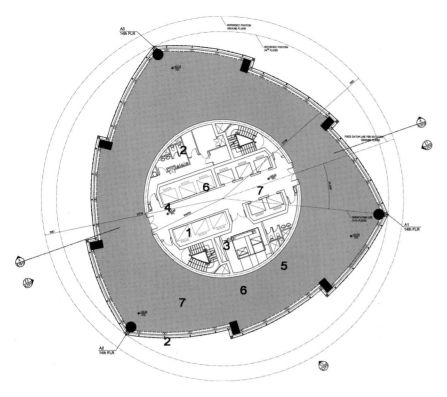

14TH LEVEL PLAN

LEGEND

1. CORE
2. FLOOR PLATE
3. ELEVATOR LOBBY
4. MALE TOILETS
5. FEMALE TOILETS
6. SERVICES
7. FIRE STAIRS

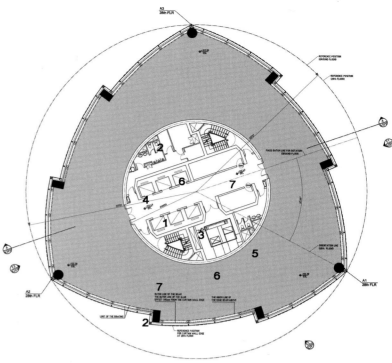

28TH LEVEL PLAN

LEGEND

1. CORE
2. FLOOR PLATE
3. ELEVATOR LOBBY
4. MALE TOILETS
5. FEMALE TOILETS
6. SERVICES
7. FIRE STAIRS

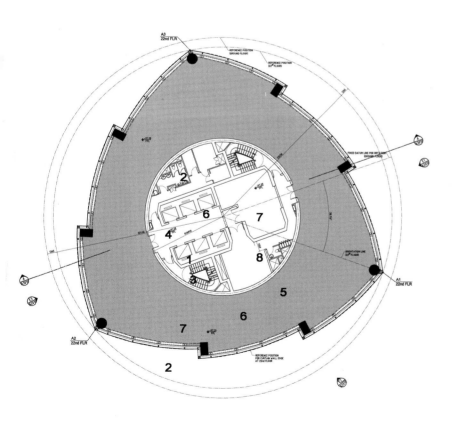

22ND LEVEL PLAN

LEGEND

1. CORE
2. FLOOR PLATE
3. ELEVATOR LOBBY
4. MALE TOILETS
5. FEMALE TOILETS
6. SERVICES
7. FIRE STAIRS
8. SERVICES

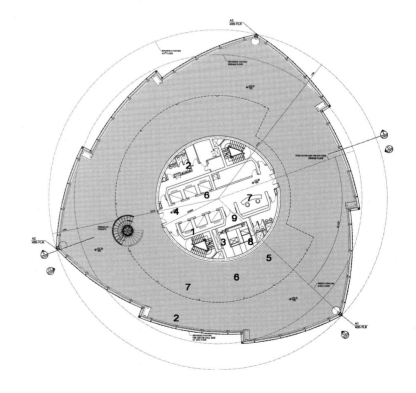

40TH LEVEL PLAN

LEGEND

1. CORE
2. FLOOR PLATE
3. ELEVATOR LOBBY
4. MALE TOILETS
5. FEMALE TOILETS
6. SERVICES
7. FIRE STAIRS
8. RECEPTION
9. UTILITY

SHANGHAI WHEELOCK SQUARE

上海会德丰广场

LOCATION:
SHANGHAI CITY,
CHINA

CLIENT: WHARF CHINA DEVELOPMENT LTD.
LEAD CONSULTANT AND PROJECT ARCHITECT: LEIGH & ORANGE LTD
ARCHITECT: KOHN PEDERSEN FOX ASSOCIATES PC
STRUCTURAL ENGINEER: AECOM ASIA CO. LTD.
BUILDING SERVICES ENGINEER: PARSON BRINCKERHOFF (ASIA) LTD.

QUANTITY SURVEYOR: DAVIS LANGDON & SEAH CHINA LTD.
MAIN CONTRACTOR: CHINA STATE CONSTRUCTION ENGINEERING CO. LTD.
AREA: 146,000 m²(GROSS STATE AREA), 12,675 m²(SITE AREA)
PHOTOGRAPHER: SHU HE

Developed by Whaf China Development Ltd., Shanghai Wheelock Square is currently the second tallest skyscraper in Puxi, Shanghai, located at the junctions of three prominent streets, namely West Nanjing Road, Huashan Road and West Yan'an Road. In this 12,675m² parcel of land, we have 4-storey basement of 33,000m², a 61-storey Grade A office tower of 107,000m², 2 detached pavilions of 5,600m² in-total for retail, food and beverage, and a piazza linking the adjacent Jing An Park.

DESIGN CONSIDERATIONS

From the first day of the project, positive response to the district green-area of Jing An Park was nailed down into the master layout design. The tower was set back from the street junction to maximize the integral piazza with Jing An Park just opposite to Huashan Road. The architecture is orientated and sculptured to address neighboring significant buildings and to capture the view of the bund. Along the high-end retail street, Nanjing Road, a sculptural retail pavilion was designed, then revised and fine-tuned to reach a balance between the Developer's design requirements, the street-scape and the expectations of the local authorities.

DESIGN AESTHETICS

A neat curtain wall in a simple form as the energy saving skin and design is the most sensible solution for the tower. A detached retail pavilion along Nanjing Road was a positive response to the streetscape and continuous shopping experience. Along Yan'an Road, there is another retail pavilion which provides prestige dinning destinations. It is also used to shield from the public the energy tower which consists of a number of centralized cooling towers and exhaust flues, chiller plants and genset at the basement levels of the pavilion away from the main tower. This prevents the Grade A office from potential disturbance by vibration and maintenance.

The process of design, submissions and construction of this project spanned over 10 years and succeeded in overcoming challenging issues such as geotechnical concern of the city with fast growing metro system, planning concerns in the territorial development of commercial street as well as increasing national concerns about energy conservation.

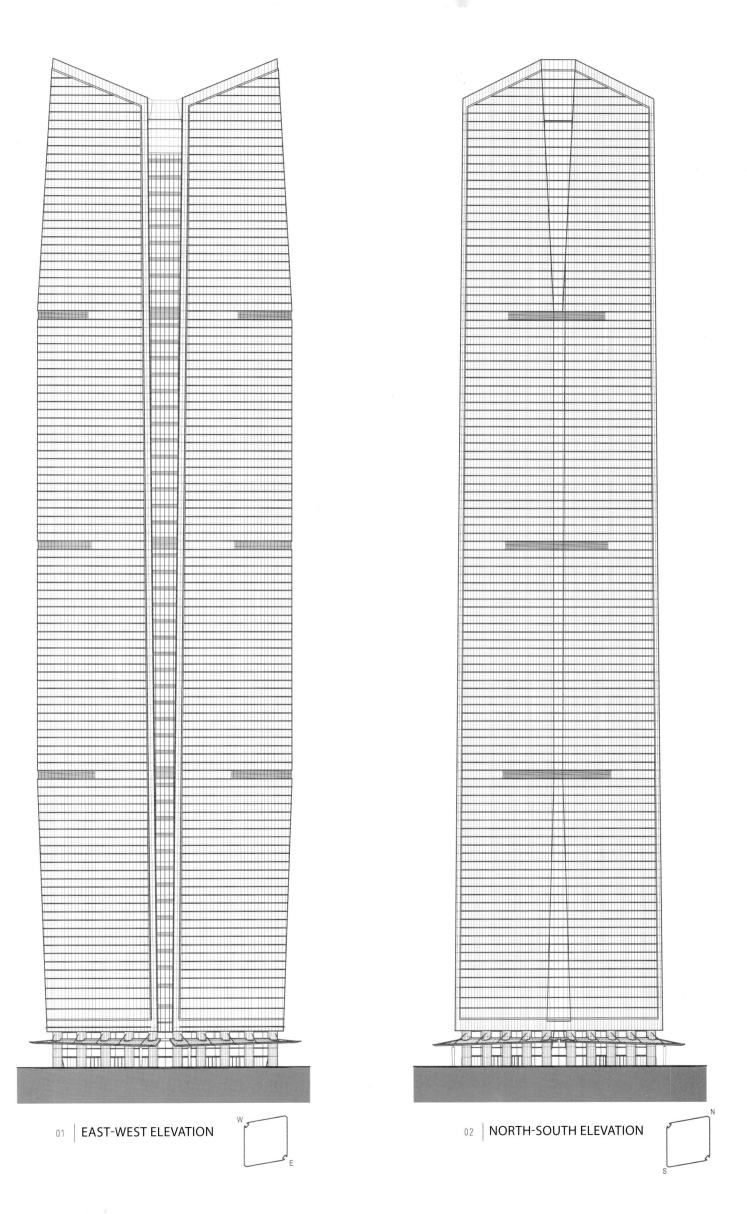

01 | EAST-WEST ELEVATION

W
E

02 | NORTH-SOUTH ELEVATION

N
S

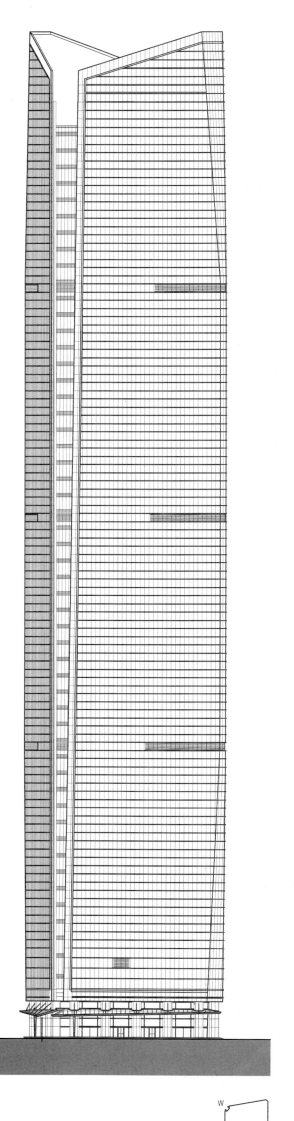

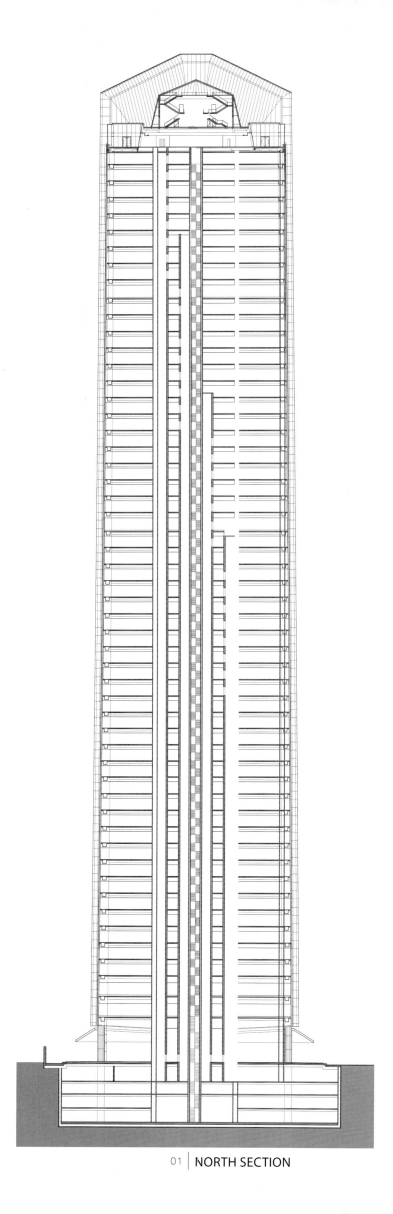

03 | NORTHEAST-SOUTHWEST ELEVATION

01 | NORTH SECTION

由中国九龙仓集团建设的上海会德丰广场是目前上海市浦西区的第二高楼，位于南京西路、华山路和延安西路三大主干道的交汇处。项目占地 12675 平方米，设四层地下室，建筑面积约 33000 平方米；地上 61 层顶级办公楼，建筑面积 107000 平方米；两个独立裙楼，建筑面积共 5600 平方米，用于零售和餐饮；此外，还有一个露天广场，与附近的静安公园相连。

设计理念

项目总体规划从一开始便确定了其要与周边静安公园的绿色环境相协调。塔楼与华山路之间留出了足够的空间来建造一片广场，与马路对面的静安公园一起营造更宽广的城市空间。建筑的朝向和装饰都在凸显其与周边重要建筑的呼应，同时还能一览外滩的景色。在高档零售街南京路上，设计了独立的零售商业裙楼，通过多番调整和论证，平衡了开发商的设计美学档次定位和当地部门对街景设计的预期。

设计美学

外观幕墙采用节能材料、设计简约，对于该建筑而言，这种设计是最明智的。南京路上的零售裙楼延续了街道景观，为市民提供了消费购物场所。在延安路上，独立餐饮裙楼还提供了高档餐饮服务。该裙楼屏蔽了后方的能源塔，有效隔离了中央制冷塔、废气管道、冷却设施和位于裙楼地下的发电机组。这一举措也免除了大型设备对顶级办公楼产生的震动和干扰。

项目从设计、报建到峻工一共持续了十多年，克服了各种挑战，包括与地铁规划发展相关的土力工程事宜，关于商业街领土的长远规划，以及国家对节能环保等方面的要求。

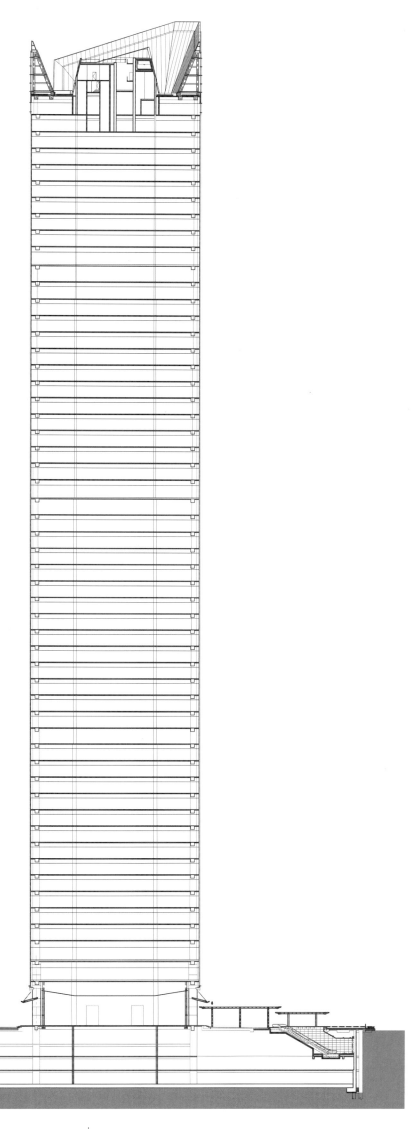

02 | **NORTHWEST SECTION**

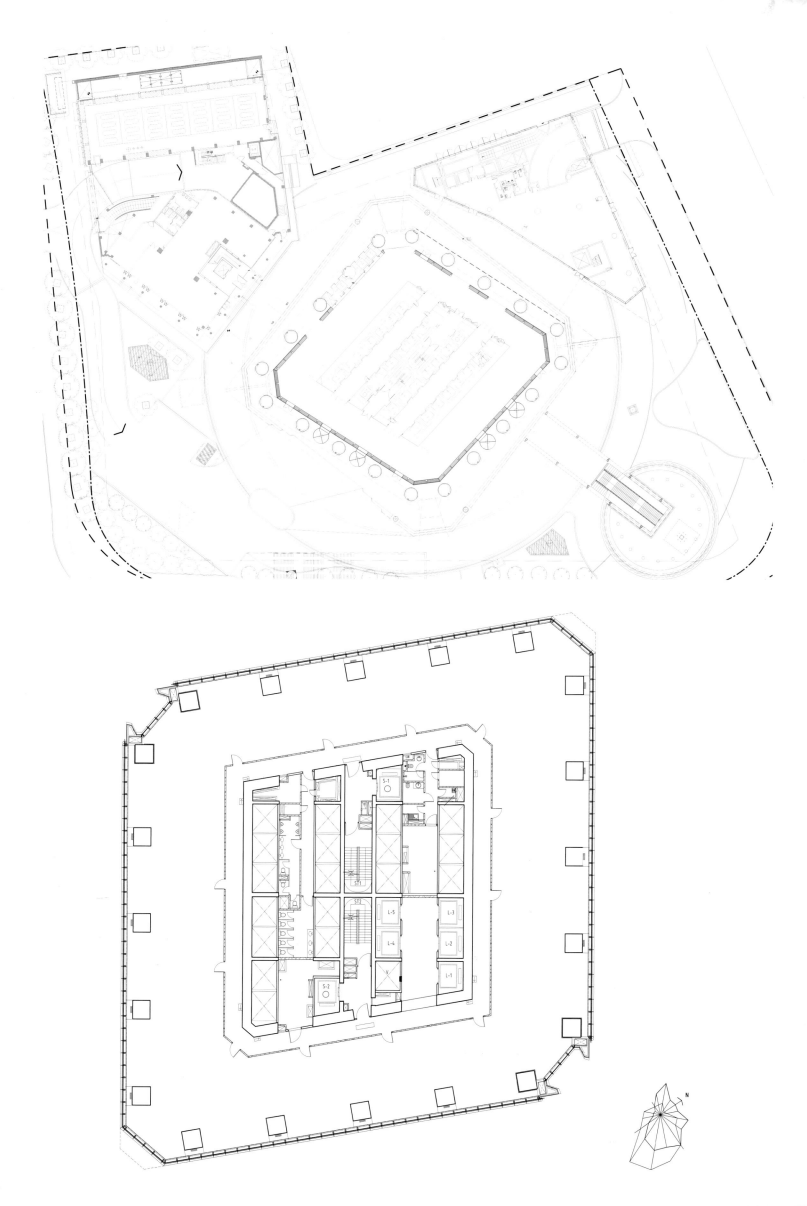

PWC TOWER

PWC 大厦

LOCATION:
MADRID, SPAIN

ARCHITECT: CARLOS RUBIO CARVAJAL, ENRIQUE ÁLVAREZ-SALA WALTHER
FIRM: R&AS
AREA: 110,000 m²

ARCHITECTUAL MANAGEMENT: JUAN JOSÉ MATEOS BERMEJO
PHOTOGRAPHER: RAFAEL VARGAS, MARK BENTLEY

The building is part of a set of four towers located in the north of Madrid. The clarity of the volume of each tower was considered important for the distant vision of the grouping in the skyline of Madrid. What gives the tower its character is its verticality. It was, therefore, more important the slenderness of the proportions of the volume than its height.

The building is divided into three parts separated by fissures that increase the sense of verticality in the volume. These fissures introduce light inside of the building and create the illusion of a group of vertical pieces.

A rigorous geometry of tangent circumferences from an equilateral triangle has been the instrumental tool in the design of the floor plan. In designing the floor plan, the need to accommodate hotel rooms has been a determining factor, as well as to make this strict scheme compatible with the open floor plan an office should have. At the center of the floor plan, a vertical nucleus is located where both elevators and electrical, heating, ventilation, and air-conditioning systems run throughout the building.

The skin of the building responds to the dual need to resolve technical issues and image through the use of a double wall. The reflections, shadows and transparencies make the building vibrate. An exterior glass skin, situated on the ends of the floor plan's pasarelle, helps the building appearance look uniform.

An internal facade directly solves the closing of the different uses, and is responsible for sealing and sound insulation without resorting to the usual curtain wall systems. The ventilation of the space between the two enclosures and the control of direct radiation help improve the temperature conditions in the interior space. The outer skin disappears in the lower levels of the building to make the interior visible. It begins from a cavity of black stone carved in the square and where courtyards are located to allow light and ventilation to the lower underground levels.

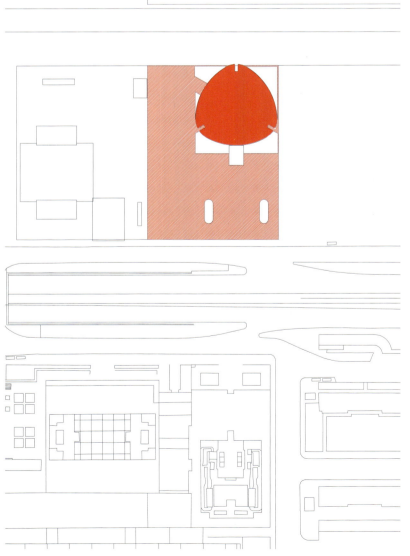

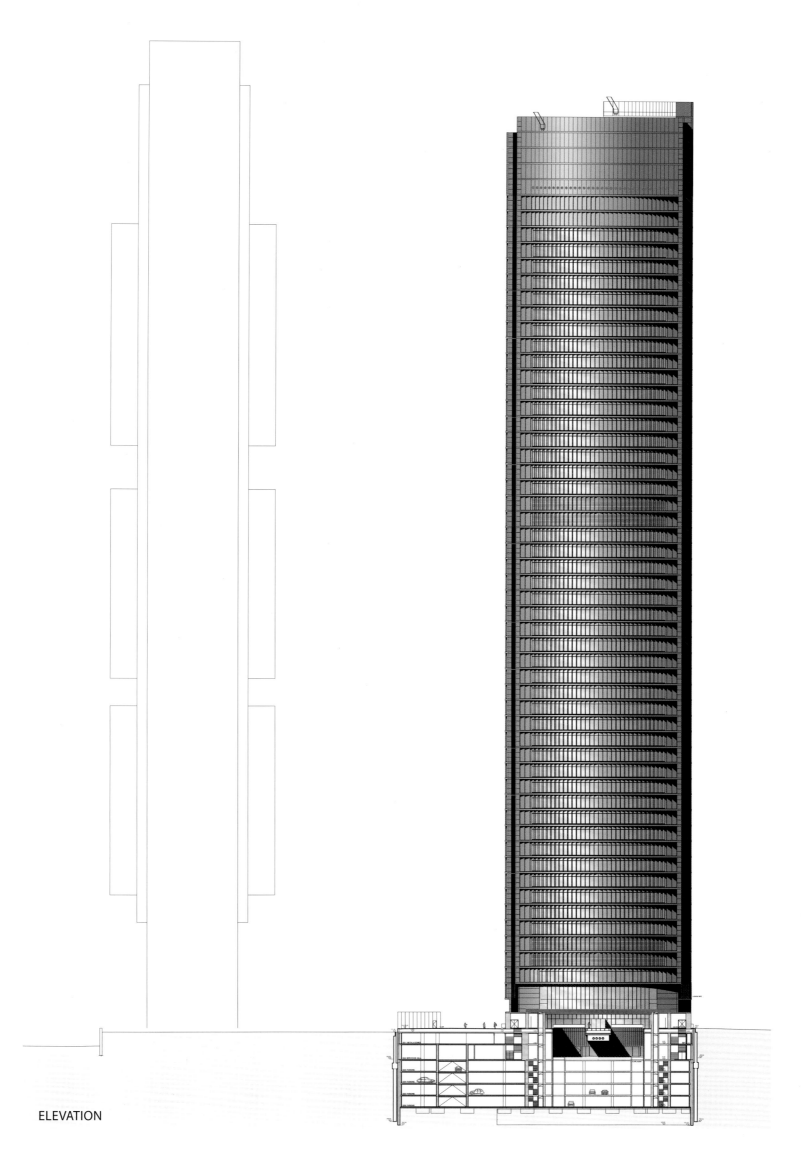

ELEVATION

大厦位于马德里北部，与周边其他三栋建筑构成一个完整的工程。每一栋大厦都层次清晰，远远望去，是这座城市的轮廓重要的一部分。PWC 大厦的特点就在于它的垂直高度。因此，建筑比例的修长比高度更加重要。

　　大厦可以分为三部分，每部分之间都有一条间隔带，这更加增添了楼层的纵深感。这些间隔带增加了大厦内部的光照，给人一种错觉，好像建筑是三块垒起来的积木。

　　将等边三角形横切一个圆面，这种几何图形成为了大厦平面图设计的重要工具。在设计中，酒店房间的容量需求是一个决定性的因素，这样严密的设计也是为了满足办公室应有的开放式平面设计需求。在设计的中央，垂直的建筑轴心则聚集了整个大厦的电梯、电力系统、供热系统、通风系统和空调系统。

　　大厦的外立面在技术问题方面采用了双层幕墙，来满足上述技术方面的双重需求。玻璃中的映像、阴影和透明设计让整个建筑动感十足。位于首层平行面末端的外部玻璃让建筑外观看起来和谐统一。

　　大厦的内部幕墙具有多种功能，具有良好的密封和绝缘效果，因而设计没有采用常见的屏式管墙。空气流通和光源控制有助于改善室内空间的温度条件。大厦底层没有设计外层幕墙，以便让室内空间一览无余。在一个黑色岩石的方格孔洞处，外墙开始出现，那里也设计了一些天井，光线和空气可通过此通道到达地下楼层。

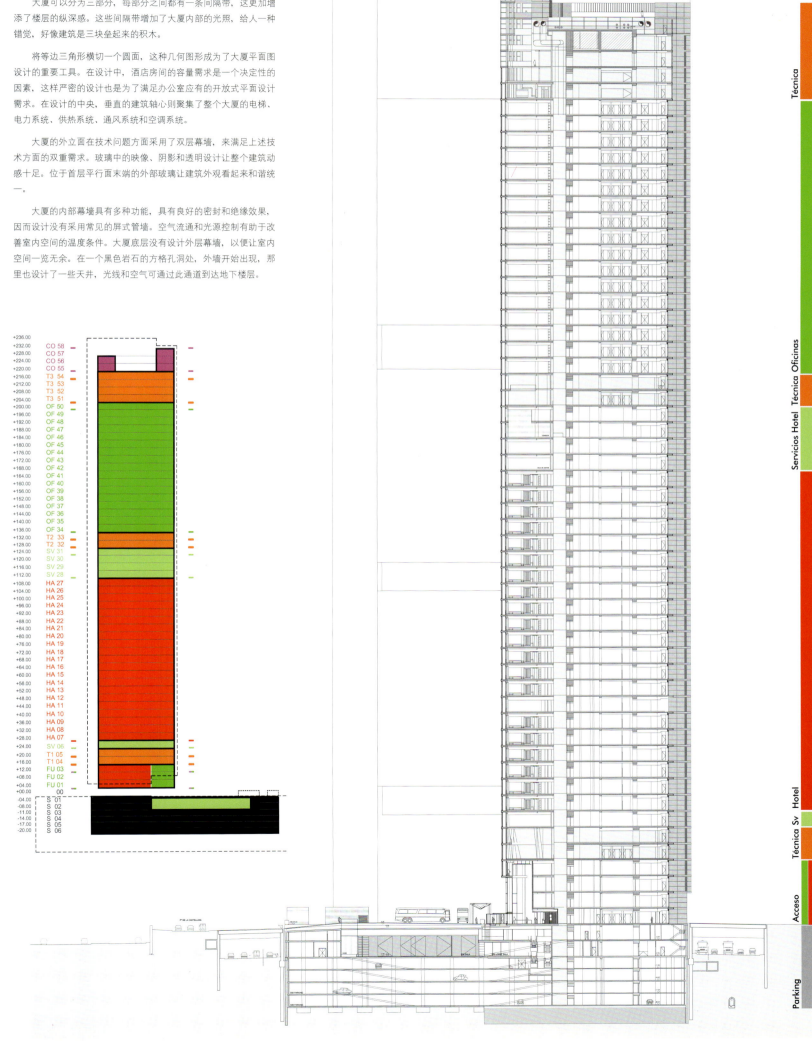

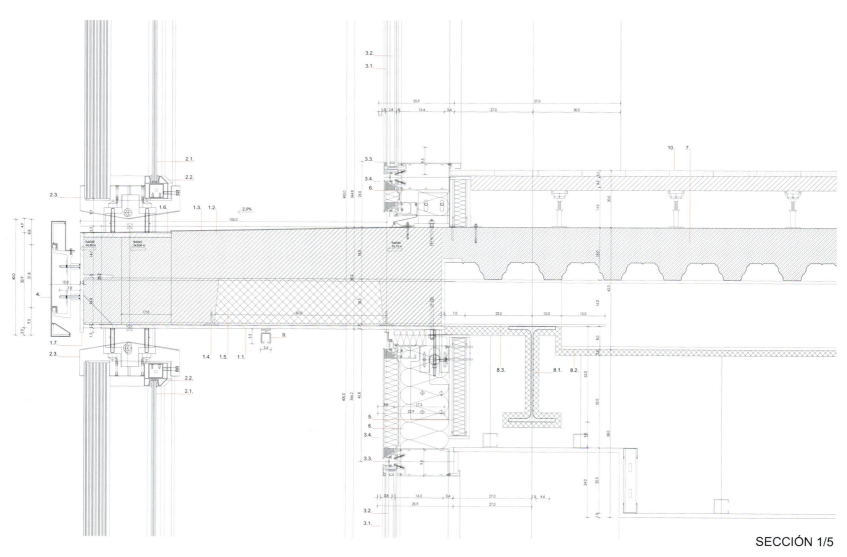

SECCIÓN 1/5

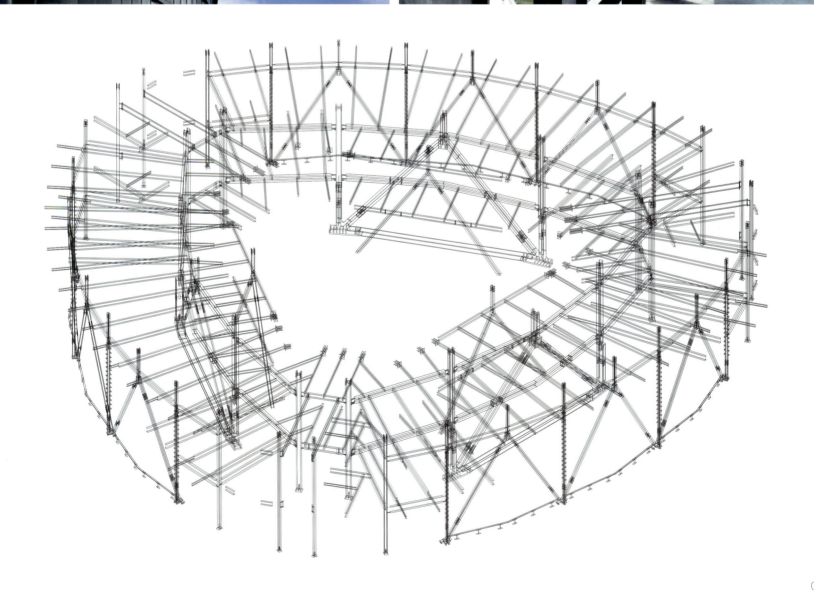

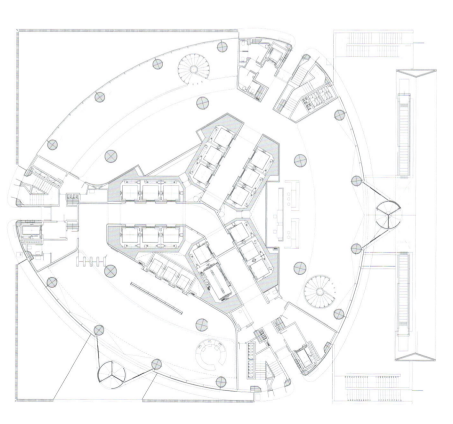

GROUND FLOOR

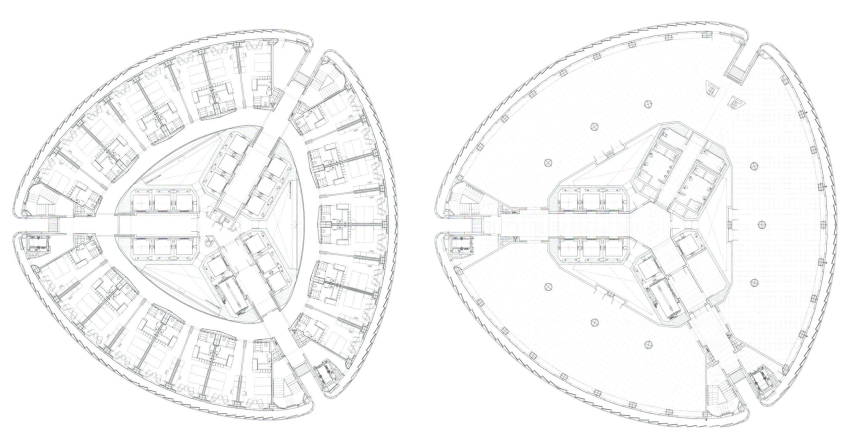

HOTEL-LEVEL

OFFICE-LEVEL

FINANCIAL TOWER

胡志明市金融大厦

LOCATION:
HO CHI MINH
CITY, VIETNAM

ARCHITECT: CARLOS ZAPATA STUDIO
ARCHITECT OF RECORD: AREP

AREA: 9, 500 m²
PHOTOGRAPHER: CHAPUIS TRISTAN, MURRAY ALAN

Shaped like a huge leaf curled in on itself and opening out to the sky, the tower evokes both nature and traditional architecture and objects made from assembled or woven plant material. The tower's complex geometry consists of inclined cylinders linked together by truncated cones. The sculptural quality of the tower is heightened at night by white lighting that emphasizes the sheer vertical of the west facde, the pinnacle of the tower and the underside of the helipad (helicopter platform) which juts out by 25 meters.

The glass panels on the facades incorporate screen-printed motifs (varying in density according to their position) which soften glare and provide protection from the sun. Inside, wooden louvred shutters filter the light in the upper and lower parts of the windows. This interplay of materials again echoes the traditional Vietnamese art of weaving and assembling natural materials.

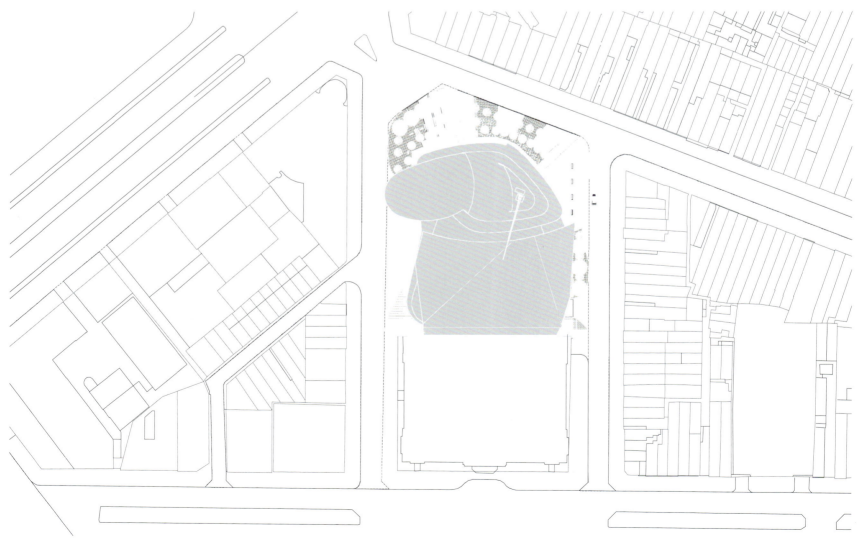

SITE PLAN

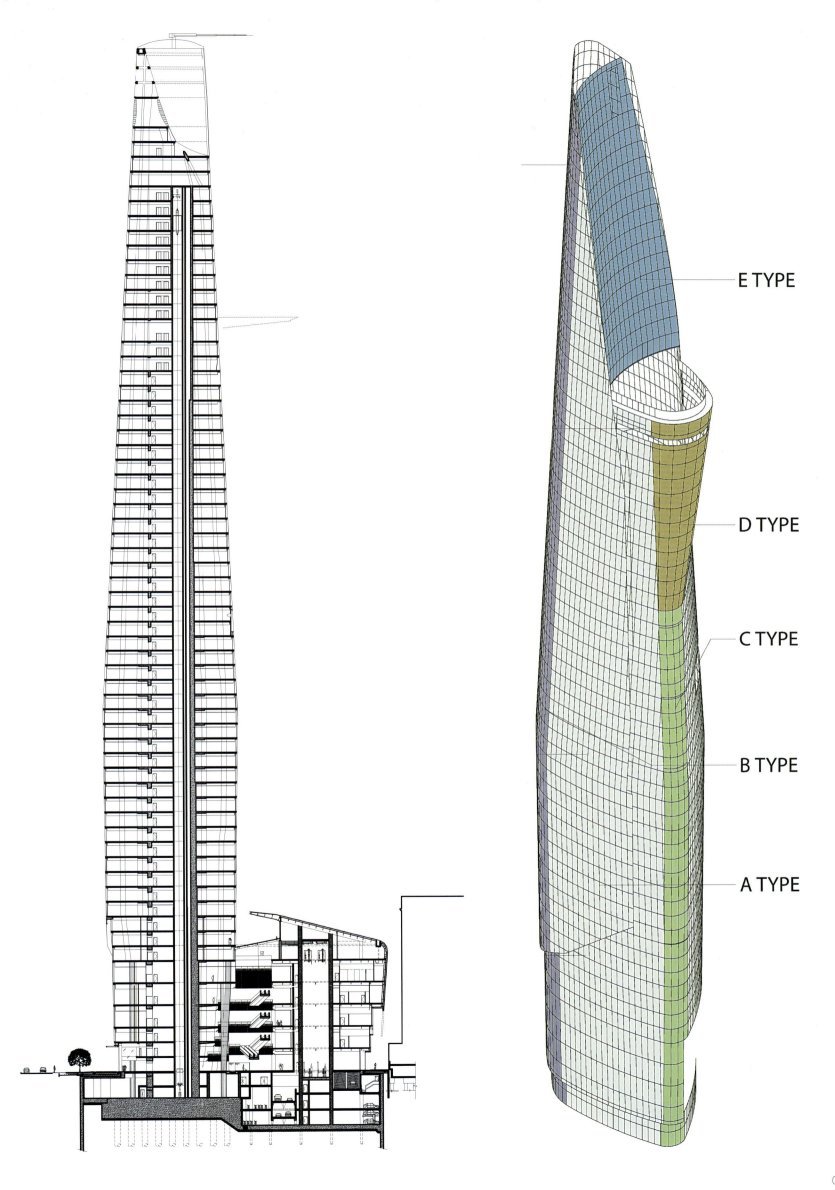

E TYPE

D TYPE

C TYPE

B TYPE

A TYPE

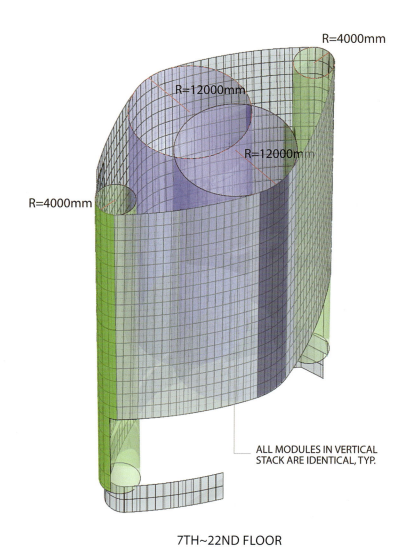

R=4000mm

R=12000mm

R=12000mm

R=4000mm

ALL MODULES IN VERTICAL
STACK ARE IDENTICAL, TYP.

7TH~22ND FLOOR

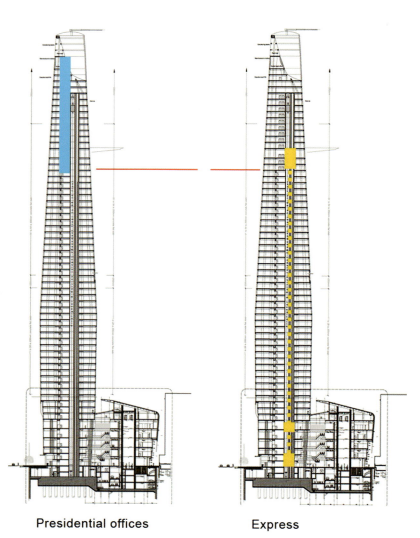

Presidential offices

Express

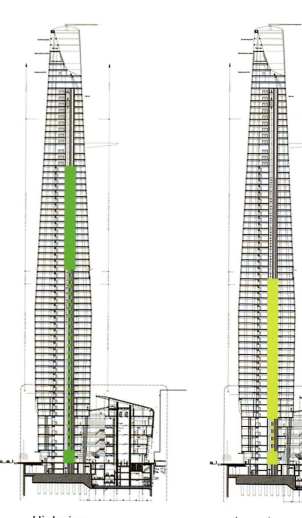

High rise

Low rise

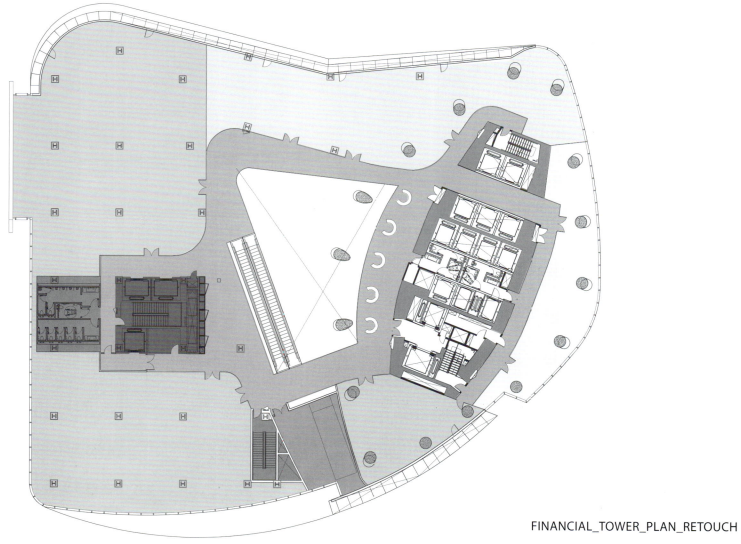

FINANCIAL_TOWER_PLAN_RETOUCH

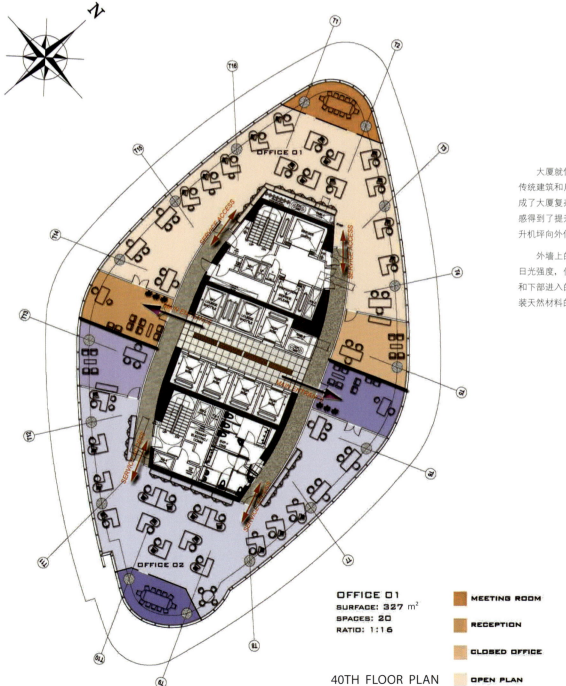

大厦就像一个巨大的叶子卷作一团，耸立于天空。该建筑同时体现了自然、传统建筑和用植物材料组装或编织的建筑的风格。斜圆柱体与截头圆锥共同组成了大厦复杂的几何形状外观。夜晚，在白色灯光的映照下，大厦雕塑般的质感得到了提升，突出了大厦西立面的绝对垂直、塔尖和直升机坪的下部。该直升机坪向外伸出 25 米。

外墙上的玻璃面板采用丝网印刷图案（位置不同，密度不同），可以削弱日光强度，使建筑免于太阳直射。大厦内部，木制百叶窗可以过滤由窗户上部和下部进入的光线。这种材料间的相互作用，再次呼应了传统的越南编织和组装天然材料的艺术。

OFFICE 01
SURFACE: 327 m²
SPACES: 20
RATIO: 1:16

- ■ MEETING ROOM
- ■ RECEPTION
- ■ CLOSED OFFICE
- ■ OPEN PLAN

OFFICE 02
SURFACE: 331 m²
SPACES: 20
RATIO: 1:16

- ■ MEETING ROOM
- ■ RECEPTION
- ■ CLOSED OFFICE
- ■ OPEN PLAN

40TH FLOOR PLAN

ALMAS TOWER

迪拜阿勒玛斯大楼

LOCATION:
DUBAI, UAE

ARCHITECT: ATKINS
CLIENT: DUBAI MULTI COMMODITIES CENTRE
AREA: 183,000 m²

Almas Tower takes the shape of two intersecting ellipses in plan, spanning 64m at maximum length and 42m at maximum width. The lower tower faces north with the taller being south facing, overlapping along their east west axis. One of the greatest challenges was dealing with the stress imbalance caused by the fact that one part of the building finishes 60m below the other. We therefore designed one end of the core thicker by 150mm and had to monitor the placement of the floor slabs. If these were not level, this would have induced more vertical stresses.

To ensure the building remained aesthetically balanced, we carried out analysis based on total vertical loading, taking into account the concrete strength, the effects of increasing the thickness of some of the core walls and the effects of creep. The final expected deflection was found to be 170mm and we did not want to deflect more than 0.001% of the building's height. The current movement has not exceeded 50mm.

The north tower has a semi-transparent elevation, thereby maximising the cool, ambient northern light; whereas the exterior of the south facing tower has a high performance finish to afford it maximum protection from the heat.

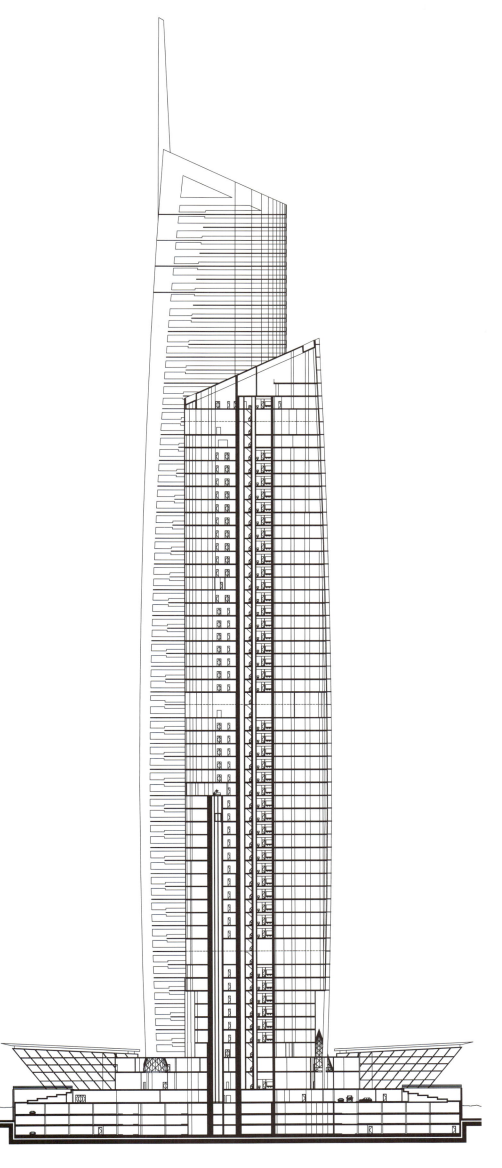

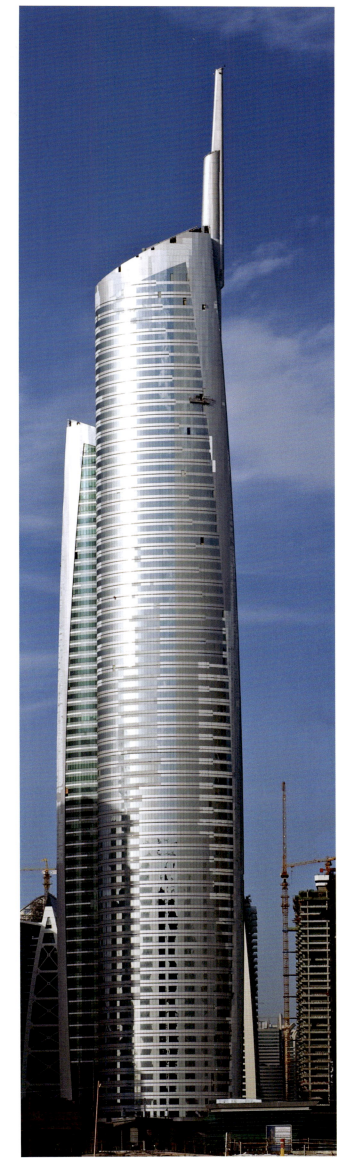

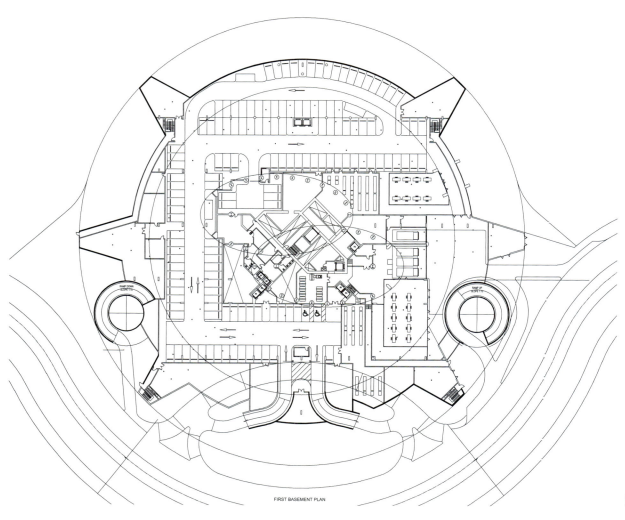

FIRST BASEMENT PLAN

FIRST BASEMENT PLAN

GROUND FLOOR PLAN

迪拜阿勒玛斯大楼的外观是两个相互交叉的椭圆面，大厦的长度横跨 64 米，最宽处达 42 米。其较低的塔楼朝北，较高的塔楼朝南，这样就使得建筑东西轴呈叠加状态。建造时，结构上最大的挑战之一是，如何处理由建筑的高低差（较高的塔楼比较低的塔楼高出 60 米）带来的压力不平衡。因此，建筑师建造的时候将核心的一端加厚了 150 毫米，并监测了楼板位置，如果不水平，会导致垂直方向上压力过大。

为了保持建筑美观协调，建筑师在分析了整体竖向荷载的基础上，同时考虑了混凝土的强度，也考虑了增加核心墙的厚度带来的影响及是否有蠕变效应。最后，最佳的偏转量是 170 毫米，建筑师希望偏转量不要超过建筑物高度的 0.001%。目前它尚未超过 50 毫米。

北塔采用半透明的外立面设计，从而可以最大限度地利用北边的侧光；朝南的塔的外立面设计则采用了多重工艺，最大限度地达到了隔热的功能。

FIRST FLOOR PLAN

SECOND FLOOR PLAN

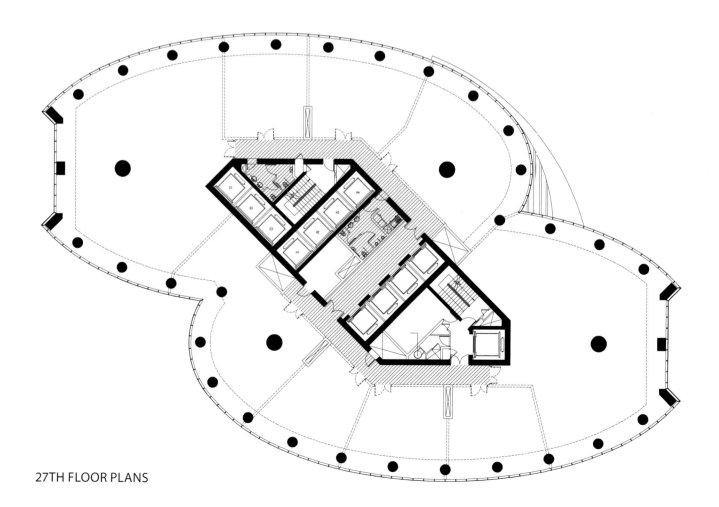

27TH FLOOR PLANS

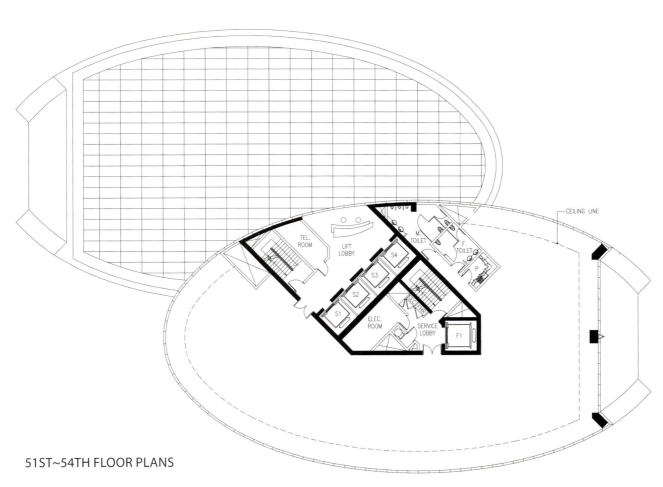

51ST~54TH FLOOR PLANS

ETIHAD TOWERS

阿提哈德大厦

LOCATION:
ABU DHABI, UNITED
ARAB EMIRATES

ARCHITECT: DBI DESIGN
PROJECT MANAGEMENT: HILL INTERNATIONAL

CLIENT: SHEIKH SUROOR PROJECTS DEPARTMENT (S.S.P.D).
AREA: 500,000 m²

The essence of the design of Etihad Towers is the harmonious integration of the unique iconic architectural form, lush indigenous landscaping and ultra-luxurious interior design, providing guests to the Jumeirah hotel with an unforgettable experience unique in the UAE and the world. Jumeirah at Etihad Towers comprises one of the five iconic towers in the Etihad Towers development, comprising a 382 room 5 star hotel and 199 serviced apartments on Tower 1 of the development –

62 levels in total.

The site of the project provides an absolutely spectacular setting with its own beachfront on the Arabian Gulf. The entire hotel – from arrival to lobby to guest room – respects this unique location with all major public spaces and guest rooms having panoramic views to the Gulf.

iconic obelisk or sculpture forms focal point of project axis

central water feature spine

oil palm arcs

curved entry wall

common area pool & children's water playground

water feature with infinity edge – 4m drop

residential towers drop-off

water feature spine

porte cochere drop-off

water curtains at entry zone

site section

0 2 5 10 20m

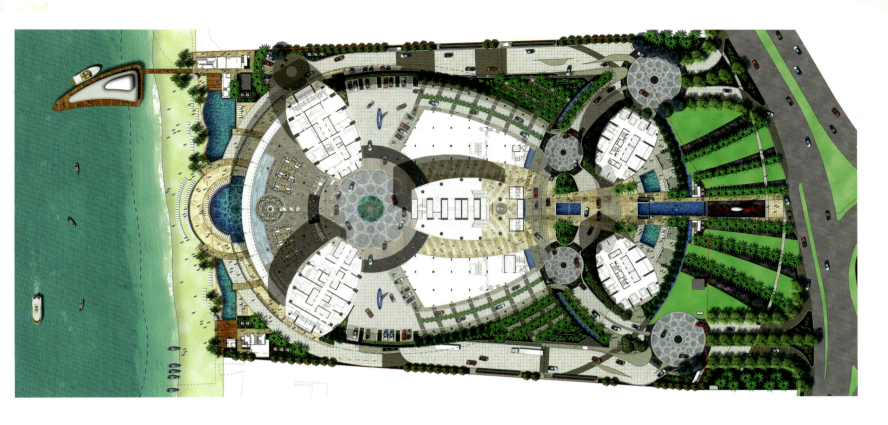

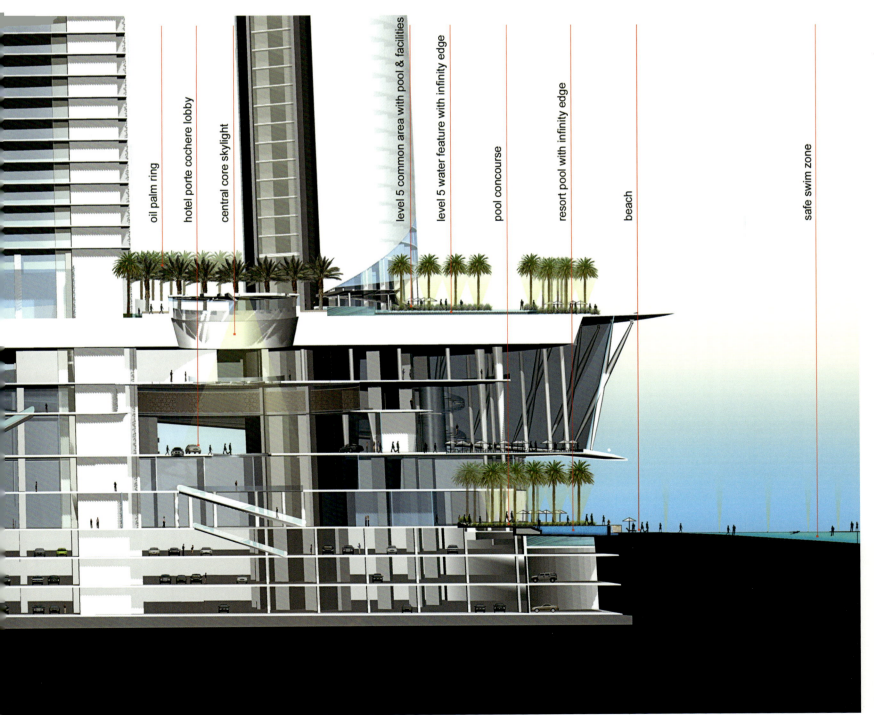

oil palm ring

hotel porte cochere lobby

central core skylight

level 5 common area with pool & facilities

level 5 water feature with infinity edge

pool concourse

resort pool with infinity edge

beach

safe swim zone

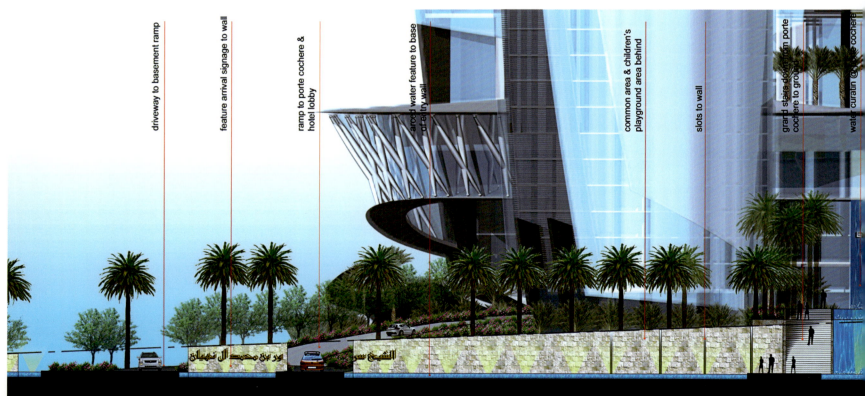

driveway to basement ramp

feature arrival signage to wall

ramp to porte cochere & hotel lobby

arced water feature to base of entry wall

common area & children's playground area behind

slots to wall

grand stair down from porte cochere to ground

water curtain @ porte cochere

arrival elevation

0 1 2   5   10m

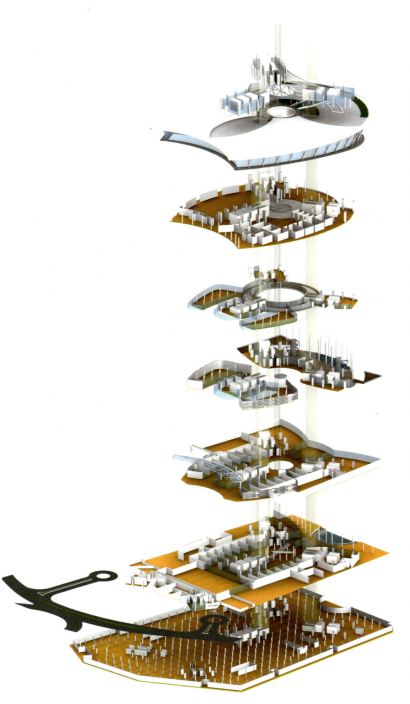

infinity edge & water curtain
from residential tower entry
level to ground

date palm avenue

stone clad curved entry wall

centre of arrival roundabout

الشيخ سرور بن   محمد آل نهيان

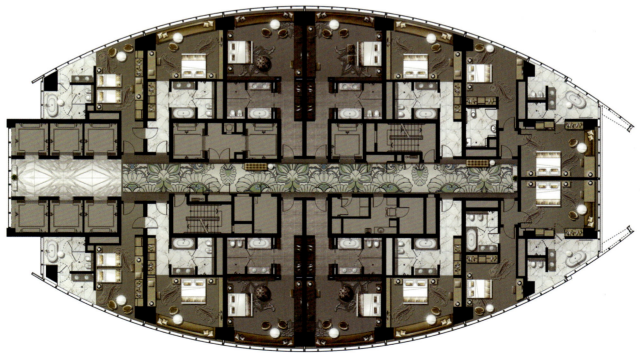

阿提哈德大厦的设计精髓，是其独特的标志性建筑形式、富有本土风情的景观以及豪华奢侈的室内设计的和谐统一，让入住卓美亚酒店的客人们享受到在阿拉伯联合酋长国乃至全球都独一无二的难忘的体验。卓美亚酒店大楼，是阿提哈德大厦五幢标志性大楼之一，在塔楼1中，有设有382间客房的五星级酒店和199间高级服务式公寓，塔楼1共62层。

阿提哈德大厦位于阿拉伯海湾的海滨地带，能饱览壮丽的海景。整幢酒店，从大厅到客房，由于其独特的位置优势，大部分公共空间和客房都能欣赏到海湾的全景。

VICTORIA TOWER

维多利亚塔

LOCATION: KISTA, SWEDEN

ARCHITECT: GERT WINGÅRDH, KAROLINA KEYZER
FIRM: WINGÅRDH ARKITEKTKONTOR AB
INTERIOR DESIGNER: ARTHUR BUCHARDT, WINGÅRDH ARKITEKTKONTOR AB, KIIL INTERIÖR

CLIENT: CALL TOWER INVEST AB
AREA: 23,000 m²
PHOTOGRAPHER: ÅKE E:SON LINDMAN, TORD-RICKARD SÖDERSTRÖM, OLA FOGELSTRÖM

As one of the tallest buildings in Stockholm, Victoria Tower serves a nearby office park with conference and office facilities, as well as providing a 229 room hotel. The tessellated facade of the building is clad entirely in colored glass, though in actuality two-thirds of the walls are fully insulated. This allows the tower to appear as a colorful prism in its geometry and materiality while meeting the needs for energy conservation. The tower has a parallelogram-shaped plan for floors 2~21, which is comprised of the hotel rooms, while the upper floors are rectangular, delineating the office spaces and a roof bar. This causes the top floors to cantilever slightly above the underlying shaft, emphasizing its geometry.

At the base of the slender tower there is a broad podium that includes a restaurant, conference area and more. The tower on top of it was in the beginning of the project a 42-story rectangle, sharpened with overhanging triangles. After several sketches and model studies, the final solution with a parallelogram that ends up with a cuboid was reached.

The highly visible position by the expressway makes the tower a literal milestone on the way toward the city center. It changes appearance with the viewer's motion, but also with the shift of daylight, weather and seasons.

The facade is made up of eight different types of glass divided so no regularities to emerge. This randomness was achieved by software developed for the project. The goal was to achieve a uniform appearance of dense and open sectors. Thin layers of metal oxides on the glass panels have been used to provide insulation and the sunscreen required. At the construction, pre-glazed modules with eight triangles each was lifted up and joined into the frame. The designer selected a glass that from the outside is rather reflective than transparent. The building is cladded in a uniform robe, a sequined dress of silver and gold, where he triangular structure strengthens the textile impression.

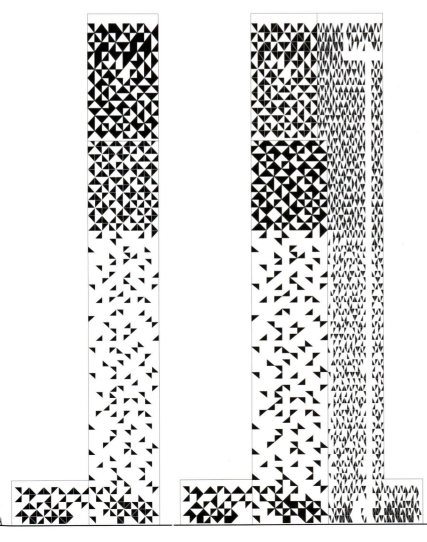

作为斯德哥尔摩最高的建筑物之一，维多利亚塔为附近的一个办公园区提供服务，包括会议室、办公设备，以及一家拥有229间客房的酒店。大楼的菱块图案的外墙面是由有色玻璃板装饰的，并且三分之二的外墙是隔热的。这样大楼看上去像是一个富有色彩的呈几何形状的棱柱，同时还满足了节能的需求。大楼第2至21层呈平行四边形，是酒店客房区域；往上是矩形，是办公空间和空中酒吧区域。这种形状使得顶层呈悬臂结构，略高于下面的支柱，达到几何效果。

细高塔楼的底部有一个宽阔的基座，设有饭店、会议室等。塔楼基座上面的部分最初的规划是建一个42层的矩形，尖锐的三角形悬挂其上。经过多次草图和模型研究后，最终的解决方案是建筑主体设计成平行四边形，顶部采用长方体。

在通往市中心的高速公路上，塔楼清晰可见，使其成为了实实在在的里程碑式建筑。观看者能随着所在位置的变换，以及日光、天气和季节的变化，看到不同的建筑外观。

塔楼立面玻璃共有8种形状，并且随意拼接。这种随意拼接是通过专门开发的计算机软件实现的，其目标是实现楼面疏密程度的统一。玻璃板上镀了一层薄薄的金属氧化物，可以起到隔热和遮阳的作用。在施工中，有8个三角形的预制玻璃模板被吊起并安装到框架上。设计者选用的玻璃，从外面看是反射玻璃，而不是透明玻璃。建筑物好像穿着一套制服，一件由金银装饰的亮片礼服，而其三角结构也加强了其"纺织品"的效果。

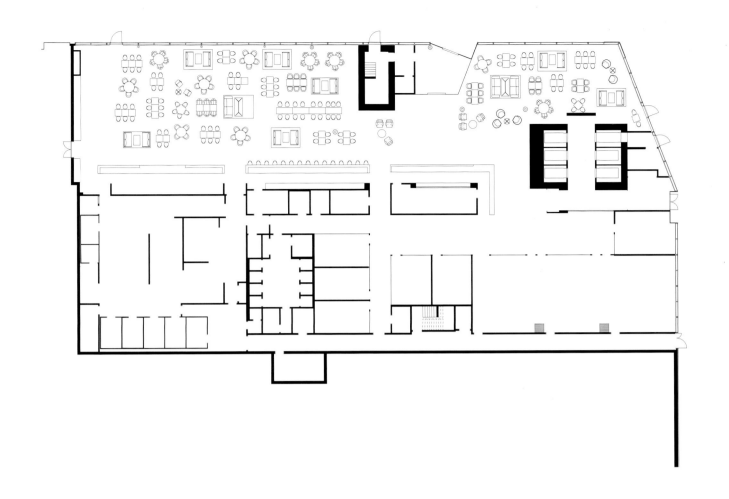

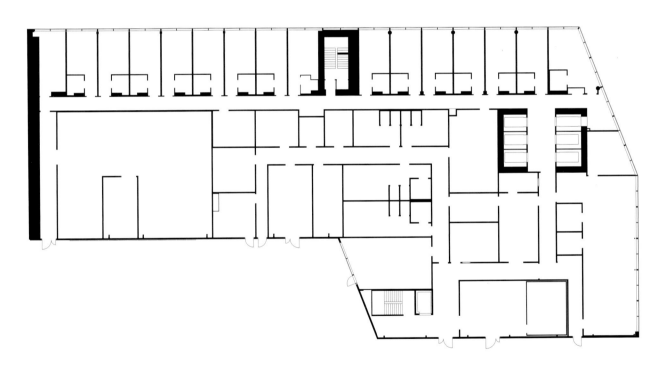

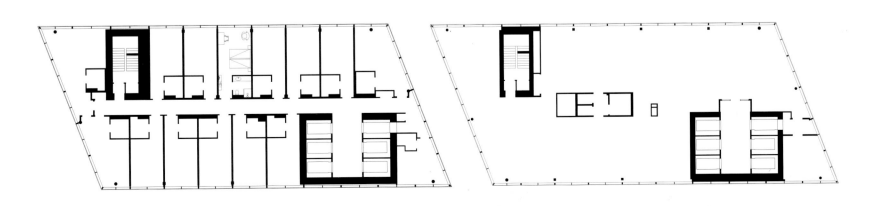

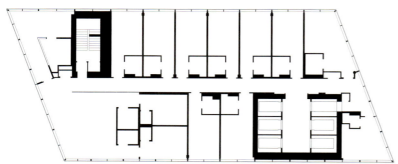

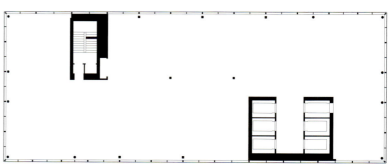

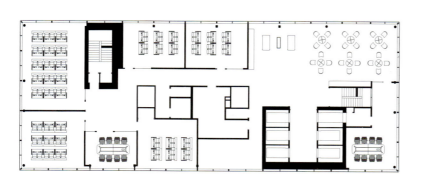

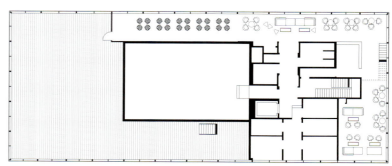

CANHIGH CENTER HANGZHOU

杭州坤和中心

LOCATION:
HANGZHOU CITY, ZHEJIANG PROVINCE, CHINA

ARCHITECT: MEINHARD VON GERKAN
FIRM: gmp Architects

COLLABORATOR: NIKOLAUS GOETZE, VOLKMAR SIEVERS
PHOTOGRAPHER: HANS GEORG ESCH, JAN SIEFKE

The "Canhigh Certer Hangzhou" is the urban link between the southern city center and the northern parts of town. A nine storey high, open and generous shopping mall acts as the new connecting street. The mall is covered by a finely worked steel and glass roof. However, the overall design – in all of its size and quality – concentrates on the 27-storey high rise at the turning of the Jing-Hang canal.

The high rise has been conceived as a large scale sculpture based on an elementary square grid of 8.40m in length. The two building volume, set back against each other according to height and length, are connected by glass joints. These two volumes of the high rise run along all of the northern and southern site boundaries thus defining all of the four corners of the square site. In the middle of the high rise

each of them step back at the entrances and, together with the battlement-like top of the towers, the high rise turns into one of the iconic landmarks of Hangzhou. All of the facades are heavily profiled and executed in grey stone. They put a special emphasis on the building's verticality and help draw the parts together as an object. Two glass bridges on level three and four link the high rise and the neighboring shopping mall.

For a new build project in China a very high standard of ecological efficiency and sustainability has been employed during the project's planning and achieved in its realization. Soon after the Canhigh Center opened it was awarded the American LEED gold certificate.

SITE PLAN

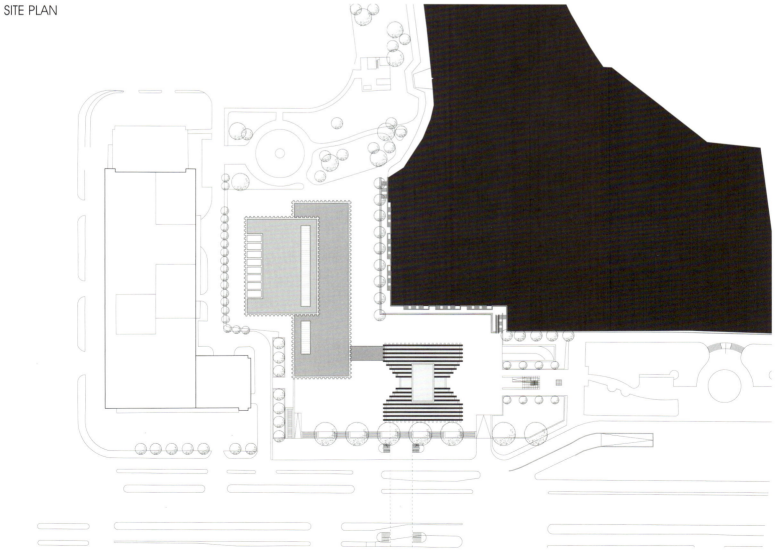

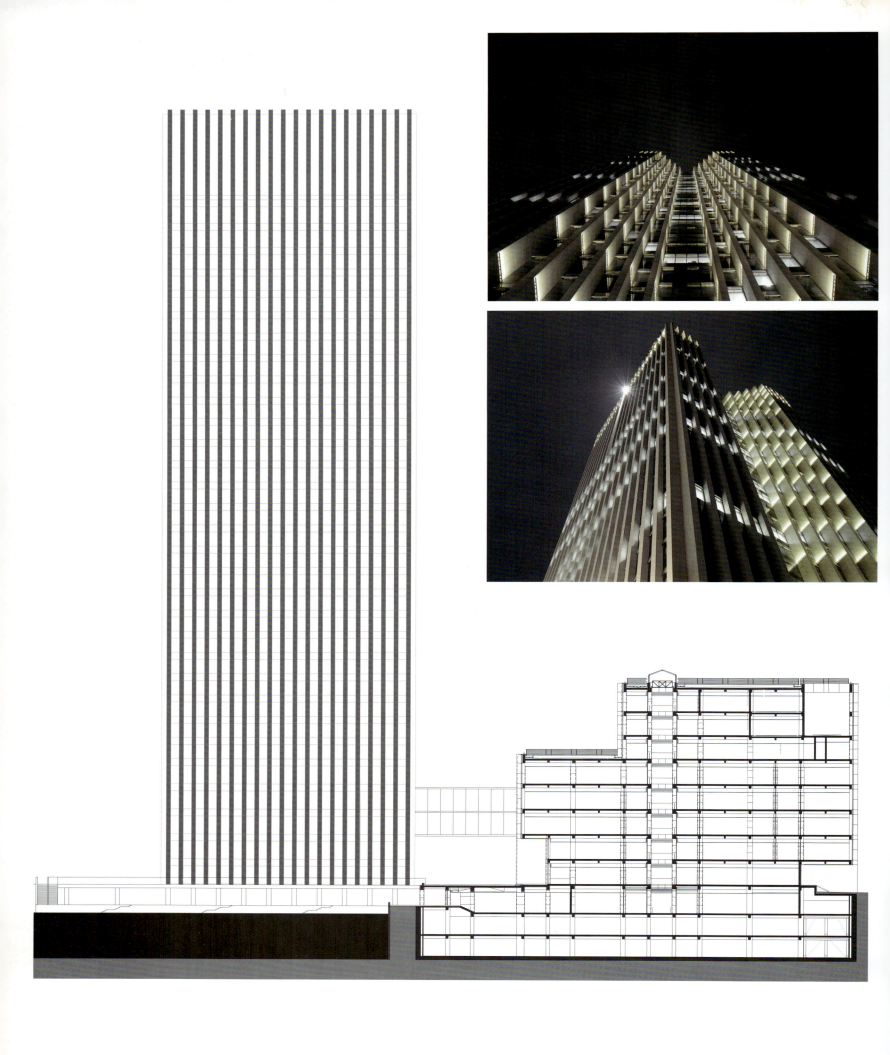

杭州坤和中心连接着南部城市中心和城市的北部地区。一个 9 层楼高、开放、宏伟的购物中心，担当了新的连接街道的角色。该购物中心由精细加工的钢结构和玻璃屋顶覆盖。然而，杭州坤和中心的总体设计，无论是规模还是质量，重点都是位于京杭大运河拐弯处的这座 27 层的高楼。

这座高楼已经被视为一座立于边长为 8.4 米的正方形基座上的宏伟雕塑。两个建筑体量，根据高度和长度相互缩进，并由玻璃接头连接。高楼的两个体量向北部和南部延伸，从而界定了正方形场地的四角。在高楼的中部，每个入口处都往里缩进，配以城墙式的塔楼顶部，使大楼成为了杭州标志性的建筑之一。所有的外墙都由石灰石塑形。重点突出建筑物的垂直度，并注重将各部分连为一个统一的建筑。在 3 楼和 4 楼有两个玻璃桥，分别用来连接高楼和邻近的购物中心。

作为中国的一个新建项目，项目规划中制定了高标准的生态效率和可持续发展性，并最终——实现。旅游集散中心在开放后不久，即获得了美国 LEED 金级认证。

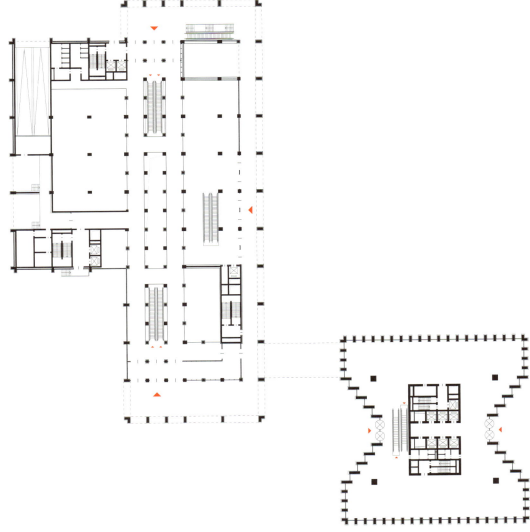

DALIAN TWIN TOWERS

百年汇豪生酒店

LOCATION:
DALIAN, LIAONING PROVINCE, CHINA

ARCHITECT: MEINHARD VON GERKAN
COLLABORATOR: NIKOLAUS GOETZE
FIRM: gmp Architects
PROJECT MANAGEMENT: DIRK HELLER, KAREN SCHROEDER

PROJECT CO-ORDINATION IN CHINA: WU DI, gmp BEIJING
CHINESE PARTNER: ECADI
CLIENT: DALIAN COMMODITY EXCHANGE, CHINA RAILWAY CONSTRUCTION ENGINEERING GROUP, PARKLAND

AREA: 353,000 m²
PHOTOGRAPHER: HANS GEORG ESCH

A central axis running from the coast to the north forms the local structural spine of the development area, with the focal point being the twin towers. On the north side, they abut the existing Exhibition Center, and rise well clear of all other buildings.

In the west and east, lower base buildings flank the twin towers, continuing the perimeter line of the semi-circular exhibition building to form a rectangular plaza, which closes off the north-facing layout of the exhibition building across the road.

Pools and green areas structure the public plaza area, and continue between the towers as an extension of the central axis as far as the adjacent pedestrian zone. With 53 above-ground storeys, the twin towers reach a height of over 240 m. Recesses in the middle, which with their large areas of glazing are reminiscent of winter gardens, are distinctive features of the basically square ground plan of the towers. The static structure of both towers consists of a reinforced concrete core plus an external grid shell, which also determines the design of the facades.

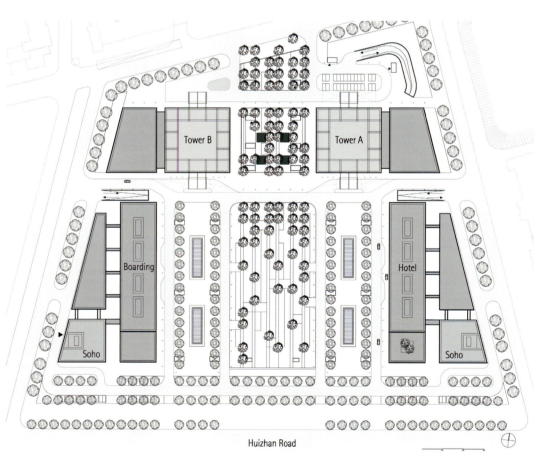

Huizhan Road

从海岸向北延伸的中轴线构成了当地开发区的建筑轴心。在这条轴心上，百年汇豪生酒店无疑是核心建筑。它的北边紧邻展览中心，高度也足以俯瞰周围的建筑。

东西方向的低层裙楼紧挨着百年汇豪生酒店，与半圆形的展览中心一起合围成了一个长方形的广场，与马路对面北向的展览中心形成隔离带。广场公共区域设有水池和绿化带，绿化带穿过两栋塔楼，沿着中轴线一直延伸，将广场与附近的步行区连接起来。

百年汇豪生酒店地上楼层达到了 53 层，总高度达到 240 米。大厦外立面中部向里凹下，采用大面积的玻璃幕墙，彰显"冬日花园"风格，成为百年汇豪生酒店这座近乎正方形的平面设计的突出特色。百年汇豪生酒店的静态结构包括一个钢筋混凝土芯和一个网状外壳，形成了建筑外观的设计。

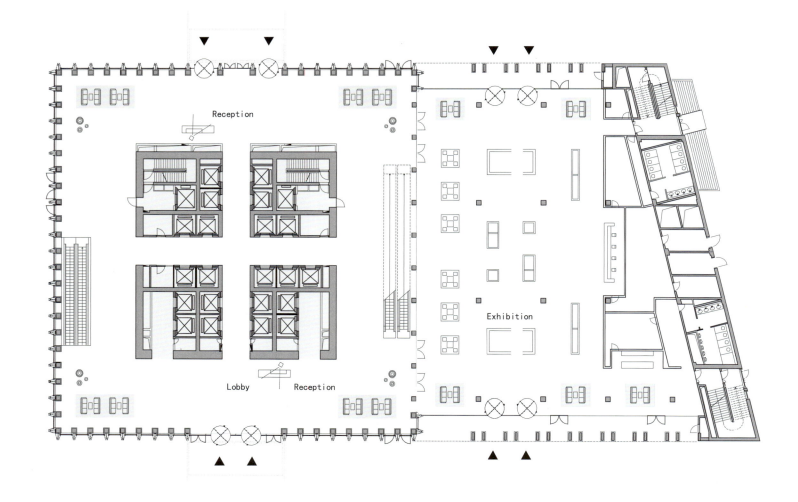

Reception

Exhibition

Lobby   Reception

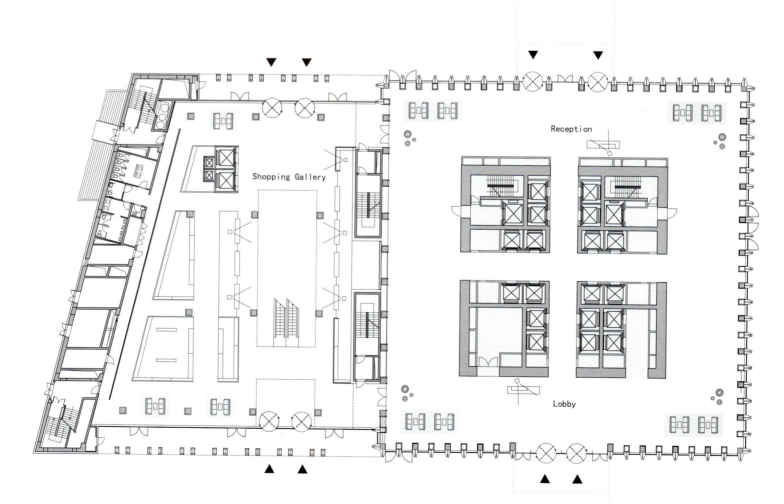

Shopping Gallery

Reception

Lobby

TOWER 185

LOCATION:
FRANKFURT AM
MAIN, GERMANY

ARCHITECT: PROF. CHRISTOPH MÄCKLER ARCHITEKTEN
CLIENT: CA IMMO DEUTSCHLAND GMBH, FRANKFURT AM MAIN
PHOTOGRAPHS: COURTESY OF CA IMMO

A new landmark has joined the skyline of Frankfurt: Tower 185 forms a gateway to the newly emerging Europaviertel district in Frankfurt am Main thus, as Germany's fourth-tallest office high-rise, extending the silhouette of the city. Its perimeter block structure references the typical shapes and materials of neighbouring buildings: beige natural stone and slate roof. The tower greets its guests from Friedrich-Ebert-Anlage with open arms. Its large drop-shaped plaza and surrounding arcades with shops and restaurants guide visitors to the entrance and invite them to linger.

SITE PLAN

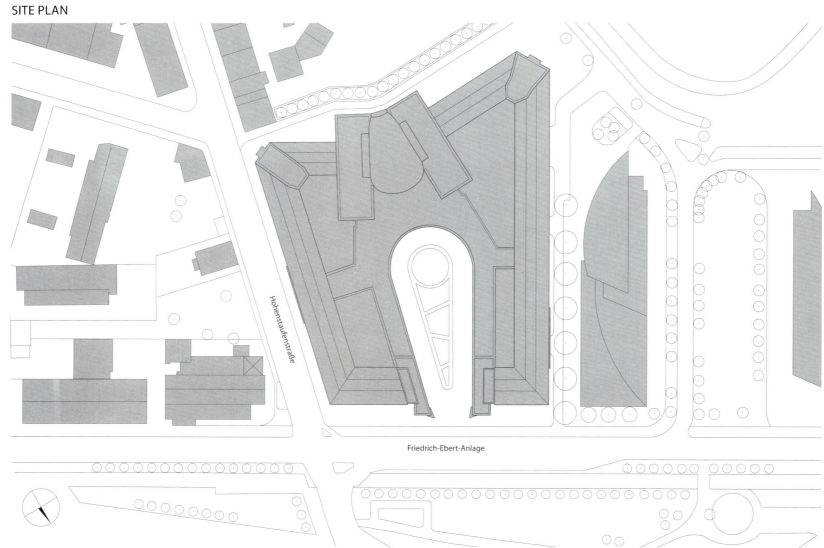

Hohenstaufenstraße

Friedrich-Ebert-Anlage

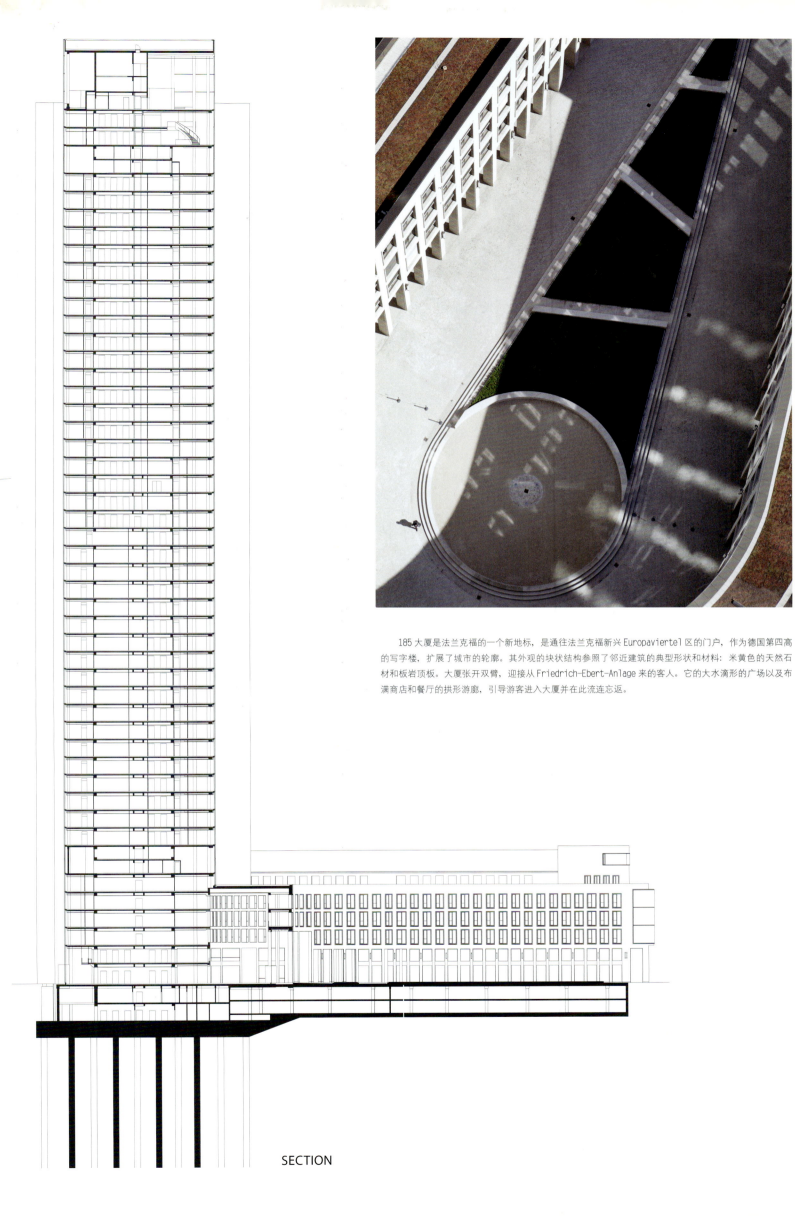

185 大厦是法兰克福的一个新地标，是通往法兰克福新兴 Europaviertel 区的门户，作为德国第四高的写字楼，扩展了城市的轮廓。其外观的块状结构参照了邻近建筑的典型形状和材料：米黄色的天然石材和板岩顶板。大厦张开双臂，迎接从 Friedrich-Ebert-Anlage 来的客人。它的大水滴形的广场以及布满商店和餐厅的拱形游廊，引导游客进入大厦并在此流连忘返。

SECTION

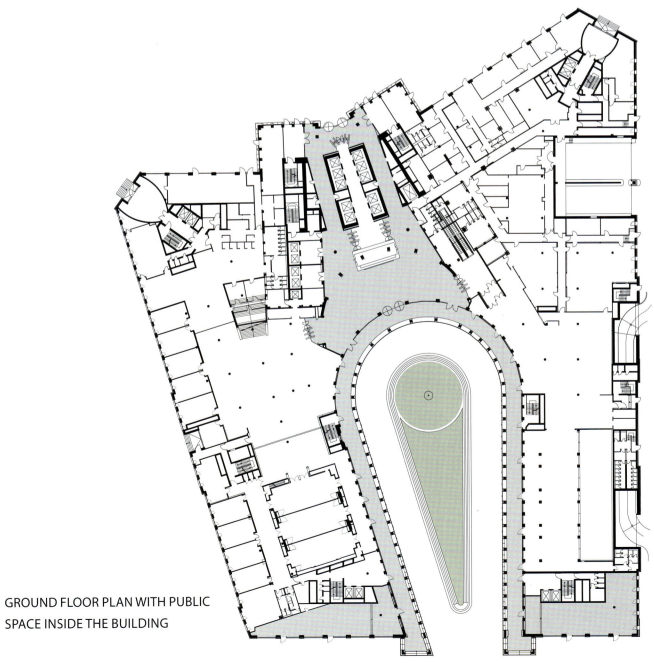

GROUND FLOOR PLAN WITH PUBLIC
SPACE INSIDE THE BUILDING

OPERNTURM

法兰克福歌剧塔

LOCATION: FRANKFURT AM MAIN, GERMANY

ARCHITECT: PROF. CHRISTOPH MÄCKLER ARCHITEKTEN
PROJECT DEVELOPER: TISHMAN SPEYER PROPERTIES DEUTSCHLAND GMBH

CLIENT: OPERNPLATZ PROPERTY HOLDINGS GMBH & CO. KG
PHOTOGRAPHER: THOMAS EICKEN, KLAUS HELBIG

The development of OpernTurm gave the city of Frankfurt the great opportunity to return to the prestigious 19th century square its formerly enclosing function as an urban ensemble. This effect was enhanced by the facades lining the square which were all done in the same beige-yellow hue as the opera building with its yellow sandstone facade in the middle. The development concept of OpernTurm reflects such design elements. Drawing on a typical urban style element of the 19th century, the perimeter development consists of two-level arcades with shops and restaurants, which gives the western side of Opernplatz its former vitality back.

Reinforced concrete and composite steel-concrete construction was used for the loadbearing structure. The floor loads are carried by the central service core and columns integrated into the facade. This means there are absolutely no columns in the interior, which results in excellent flexibility when planning the interior layout.

At the same time, more than 50% of the facade is closed, which together with the highly efficient glazing reduces the solar gains and hence the need for mechanical cooling. Compared to an all-glass facade, the stone facades of the OpernTurm project therefore save 20% of the energy normally required for cooling an office floor.

Technical services are designed to provide the highest level of user comfort and flexibility combined with low energy solutions.

©Thomas Eicken

©Thomas Eicken

© Thomas Eicken

©Thomas Eicken

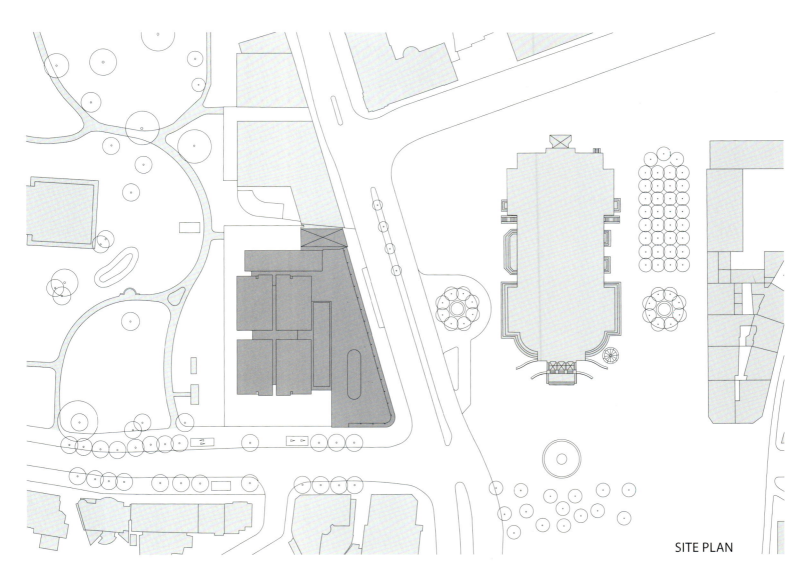

SITE PLAN

©Thomas Eicken

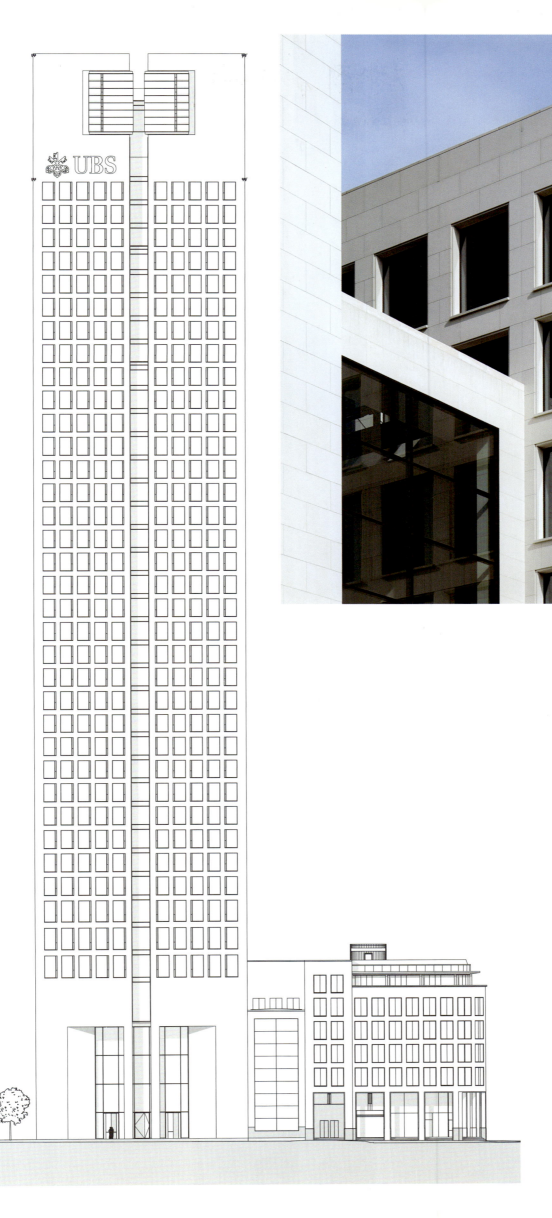

UBS

©Klaus Helbig

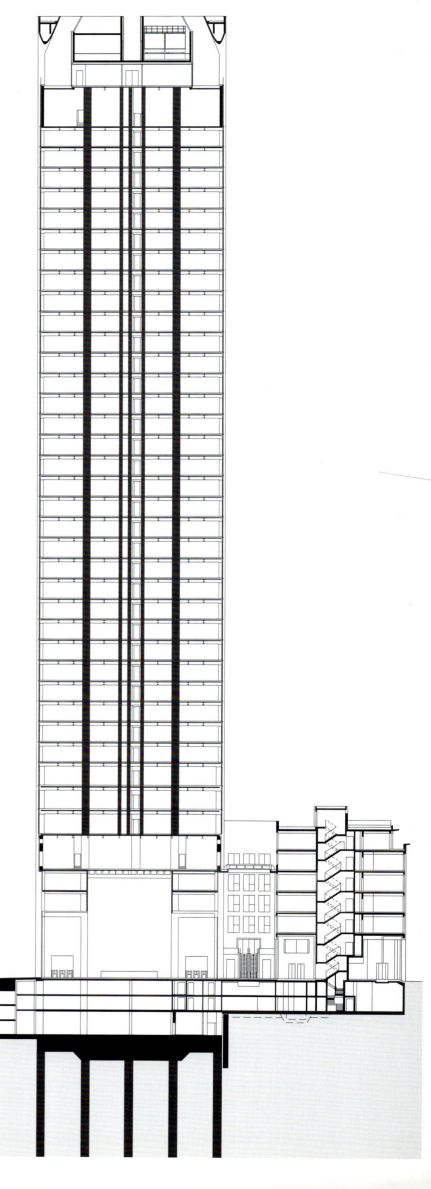

歌剧塔工程让法兰克福这座城市再次回到了 19 世纪的繁华。歌剧塔广场还原了古城魅力，展示了现代时尚。紧邻广场的外立面统一采用了米黄色的色调，与歌剧塔的中间幕墙的黄色砂岩相呼应。歌剧塔的设计理念体现了 19 世纪经典的都市风格元素，其周边包括一个两层的商场，内有各色商店和餐馆，让广场的西侧恢复了昔日的活力。

钢筋混凝土和钢－混凝土组合结构被用在了承重墙的建造中。楼层负荷由建筑的轴心柱承载，轴心柱与外立面相连。这就意味着建筑的内部没有一根柱子，大大地提高了室内设计的灵活度。

同时，50% 以上的外立面呈封闭式结构，再加上高效能的玻璃，降低了太阳光照的强度，也减少了冷却系统的使用。与全玻璃结构的外立面相比，歌剧塔的岩石外立面，降低一层办公楼的温度可同比节省 20% 的能量。

技术维护以及低碳设施为用户提供了最优的舒适度和灵活度。

©Klaus Helbig

©Klaus Helbig

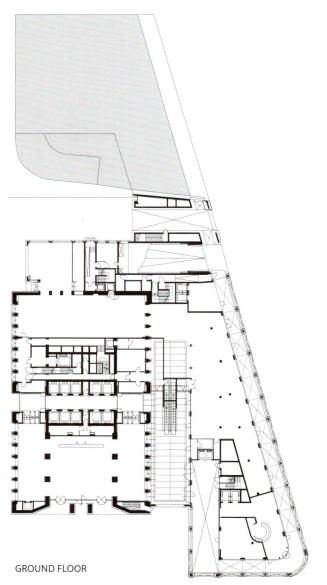

GROUND FLOOR

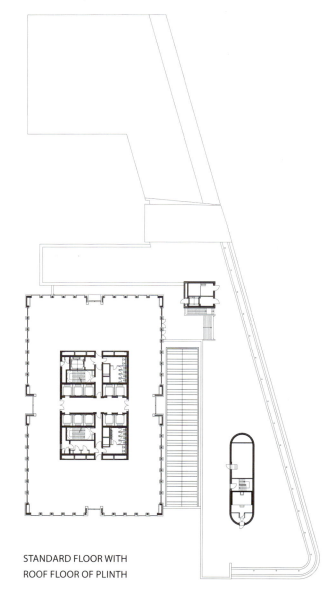

STANDARD FLOOR WITH
ROOF FLOOR OF PLINTH

©Klaus Helbig

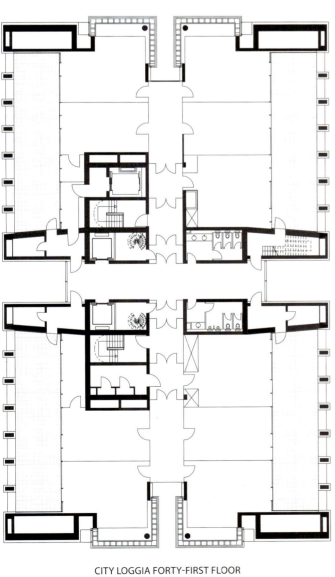

CITY LOGGIA FORTY-FIRST FLOOR

brot&butter

©Thomas Eicken

NABEREZHNAYA OFFICE TOWERS

湖畔大厦

LOCATION: MOSCOW, RUSSIAN FEDERATION

ARCHITECT: RTKL ASSOCIATES INC.
CLIENT: ENKA INSAAT VE SANAYI A.S.

PHOTOGRAPHER: DAVID WHITCOMB, COURTESY OF ENKA

Innovatively combining modern design and comfortable, functional space, Naberezhnaya Office Towers form a contemporary working environment that sets a new standard for today's Russian business centers.

ENKA, one of Moscow's most prestigious developers, set out to capitalize on 200,000m² of rentable area in the heart of the city. The resulting project is comprised of three Class-A office towers of 17, 27 and 57 storeys respectively. The towers feature high ceilings, flexible and open floor plates, and internal partitioning tailored to tenant requirements. Underground parking and a sound-insulated facade adapt to the challenges of the urban setting, and the varying building heights allow for spectacular views of the city center and the Moscow River. Occupants of the office buildings will enjoy convenient access to a transport system, a shopping mall and upscale residential options, making the facility not only a premium workplace but an integral element of Moscow's urban landscape.

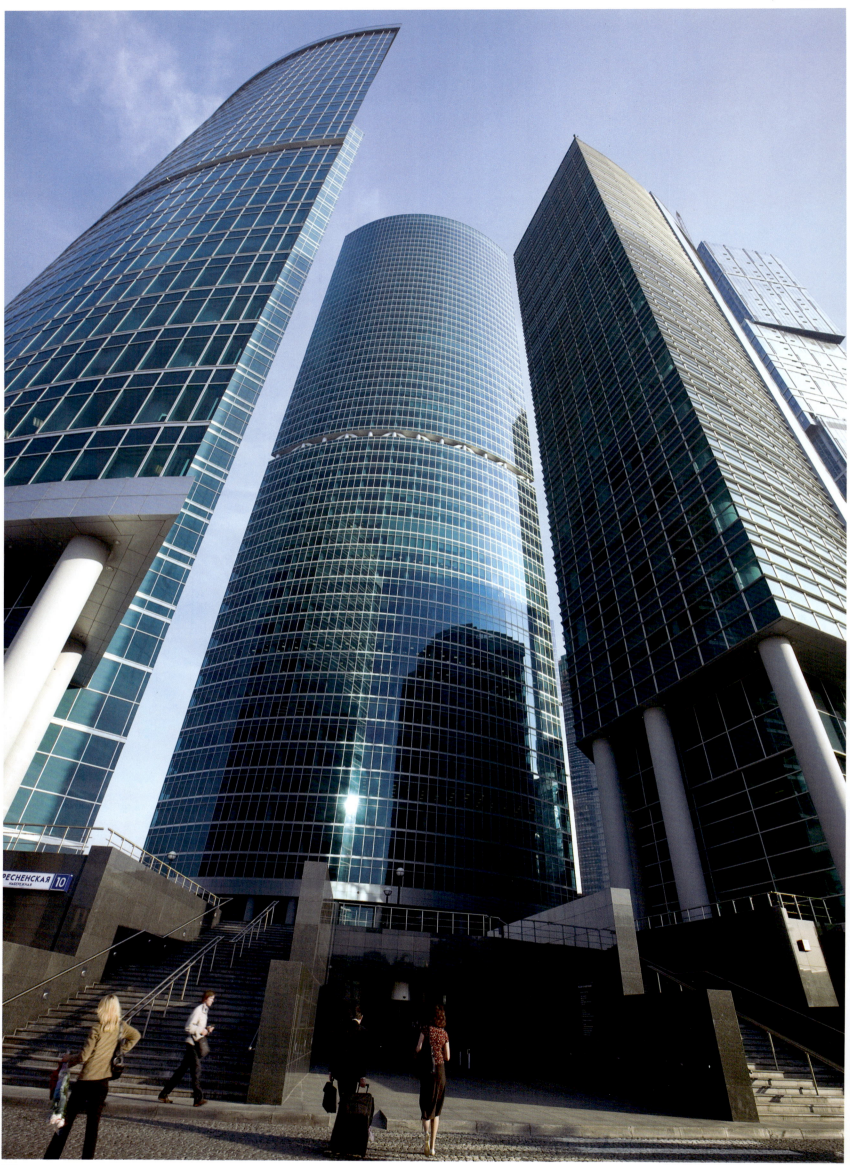

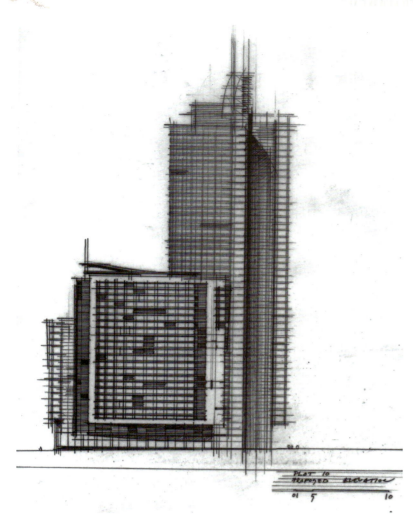

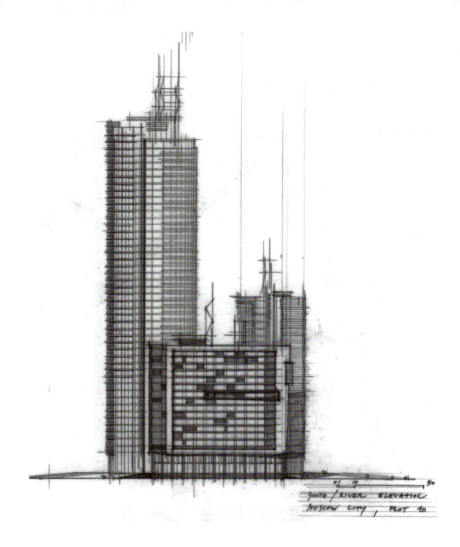

湖畔大厦独具匠心地将现代设计和舒适、多功能的空间结合在一起，营造了一种现代化的工作环境，为现今的俄罗斯商业中心设计设定了全新的标准。

ENKA，莫斯科最负盛名的开发商之一，准备在莫斯科市中心投资建造一个200000平方米的出租区域。建成后，整个项目包括三座A级办公楼，分别有17层、27层和57层。各栋大楼的吊顶甚高，楼板灵活开阔，内部空间可按租户要求分割。地下停车场和隔音幕墙能适应都市环境的挑战。站在错落有致的建筑物上，可以尽情领略市中心和莫斯科河的壮丽景观。办公楼内人员还可通过快捷通道进入交通系统、大型购物商场和各类高端住宅。这使得该项目不但成为了一个绝佳的工作场所，也成为了莫斯科城市风景中不可分割的组成元素。

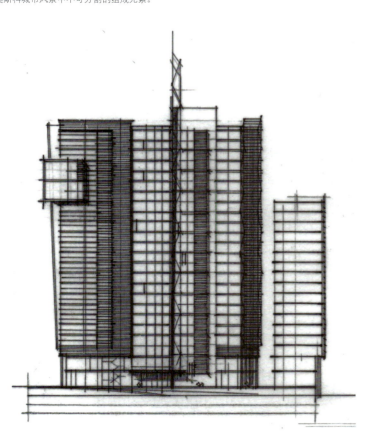

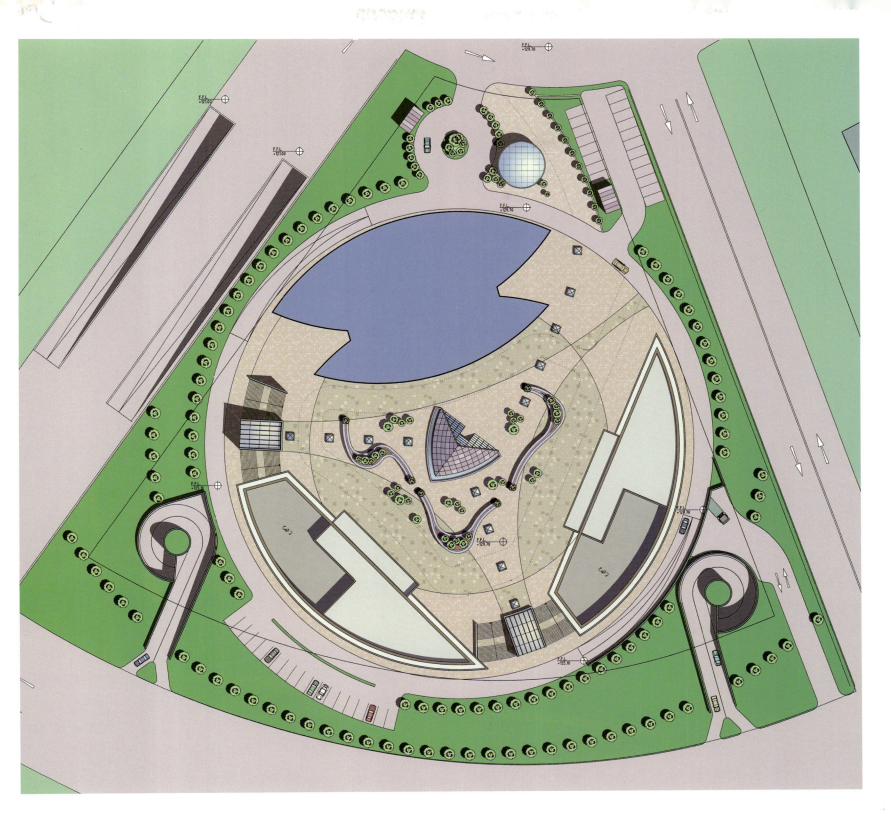

DIAGONAL ZEROZERO TELEFONICA TOWER

西班牙电信大厦

LOCATION:
BARCELONA, SPAIN

ARCHITECT: EMBA_ESTUDI MASSIP-BOSCH ARQUITECTES
PROJECT PRINCIPLE AND DIRECTOR: ENRIC MASSIP-BOSCH, ALEIX ANTILLACH, ELENA GUIM, JON AJANGUIZ
PROJECT TEAM: ESTEVE SOLÀ, RICARDO MAURICIO,

CARLOS CACHÓN, CORNELIA MEMM, CRISTINA FEIJOO, HEIDI REICHENBACHER, RITA PACHECO, RODRIGO VARGAS, JANA ALONSO, MARTA MARCET, MARIANA ARÁMBURU
DEVELOPER: CONSORCI DE LA ZONA FRANCA DE BARCELONA

AREA: 25,327 m² (ABOVE GROUND BUILT AREA),
8.576 m² (UNDERGROUND BUILT AREA)
PHOTOGRAPHER: EMBA, TAVISA,
PEDRO ANTONIO PÉREZ

The project generate a contextual tower which is at the same time a landmark and a public space, taking the urban alignments that form the perimeter of the plot as the generator of the project. It is a trapezoidal prism, sharp and stylized, a clean and serene form shimmering and light. Its transparency reveals the interior's dynamic volumes which respond to the different situations of the program, relating the tower to the varying heights of the neighboring buildings. The exterior relates to the city and the view from afar while the interior relates to the program and the close-up vision in response to the two simultaneous scales that such tall buildings must achieve.

The ground floor follows the slope of the adjacent streets and it has been developed over three levels which are open to the general public. The upper floors are open-plan office spaces, taking advantage of the structural system. This system is a modified tube-in-tube scheme.

The floors are concrete slabs that connect the perimeter structure with the central core to form a combined structural device. The bracing elements create an exterior diamond lattice that bears the stresses of each part of the building, with a higher concentration of structural elements in the lower half and a lighter density in the upper parts.

The facade is a modular curtain wall made of white aluminum profiles and extra-clear glass with white ceramic paint serigraphy on the outside. The pattern of this serigraphy follows a vertical composition that reinforces the slenderness of the tower and contributes to its changing whitish image. In combination with the inner structure, placed every 1.35 meters, and the exterior structure, this pattern also contributes to the diffusion of solar light and to glare control, generating interiors of great perceptual quality and remarkable visual comfort.

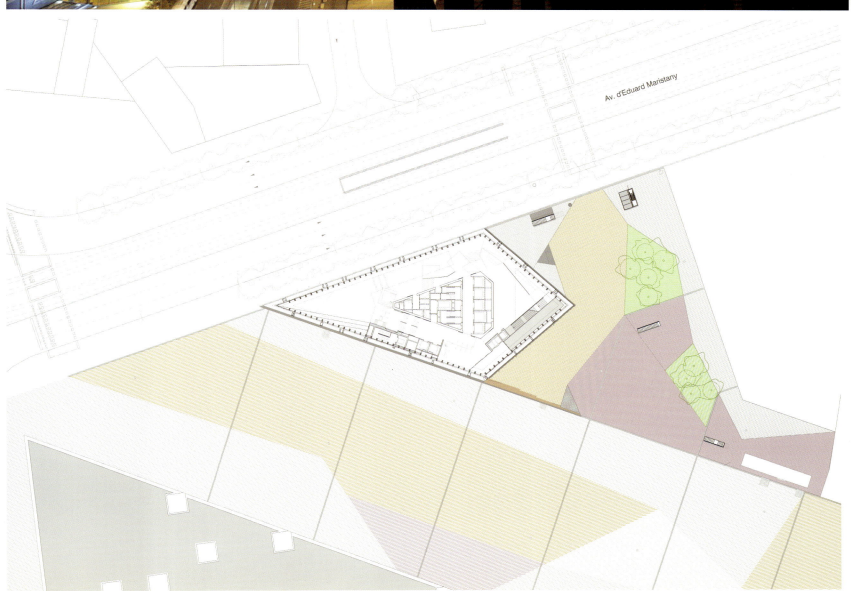

Av. d'Eduard Maristany

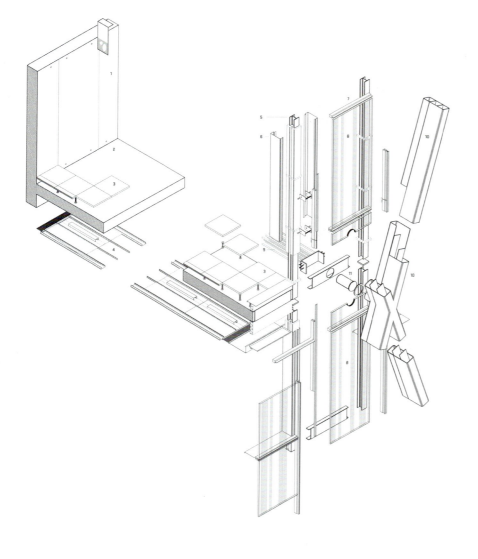

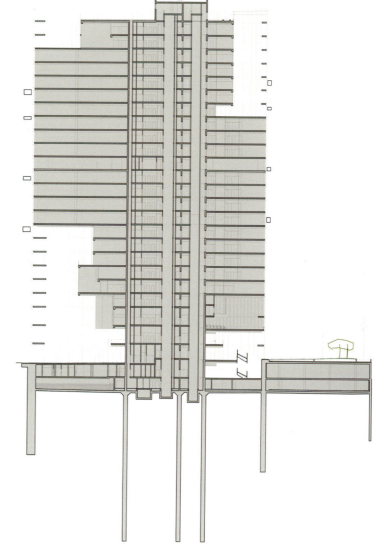

该项目意在建造这样一座大厦：它既是一个地标，又是一个公共空间，并与周边城区规划相统一。建筑呈梯形棱柱体，轮廓鲜明而风格化，干净清澈的外表熠熠生辉。它的透明度既显示出了与室内各种特色功能相呼应的动态体量，又与周围错落的建筑物相协调。建筑从外观设计上与城市形象及远处的风景遥相呼应，而室内又与各种功能及近景相得益彰。这也回应了此类高层建筑必须设法解决的两种尺度问题。

大厦底楼顺着毗邻街道的斜坡而建，底楼共三层，并且向公众开放。上部楼层为开放式办公空间，充分利用了改良后的套管式结构体系。

大厦楼层板面用的是混凝土板，通过中央支柱和周边框架连接，从而形成组合结构设计。各个结构构件形成了一种钻石晶格立面，分别承载着建筑各部分的压力，主要承重构件都集中在建筑的底部，上部较少。

大厦立面采用了模块式幕墙，由白色铝材和超级透明的玻璃构成，玻璃上印有白色陶瓷颜料绢印图案，此种图案强化了建筑的细长感和洁白的体量。图案间距为1.35米，与内部和外部结构相结合，既扩散了太阳光线，避免炫光，又创造了极佳的室内观景点。

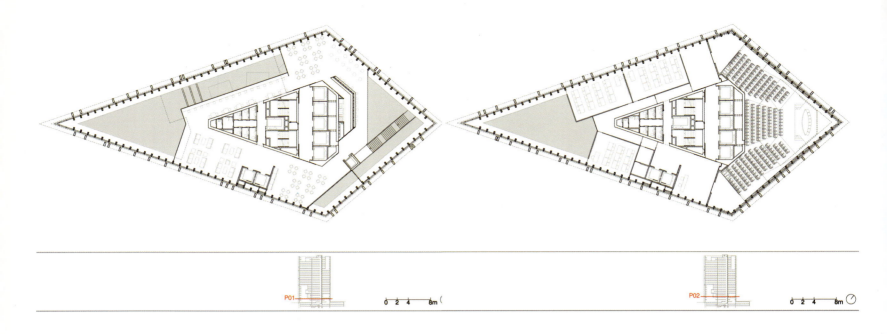

P01   0 2 4   8m

P02   0 2 4   8m

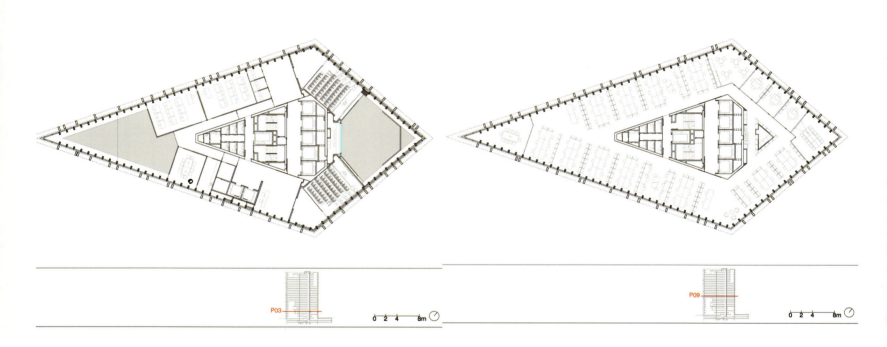

P03   0 2 4   8m

P09   0 2 4   8m

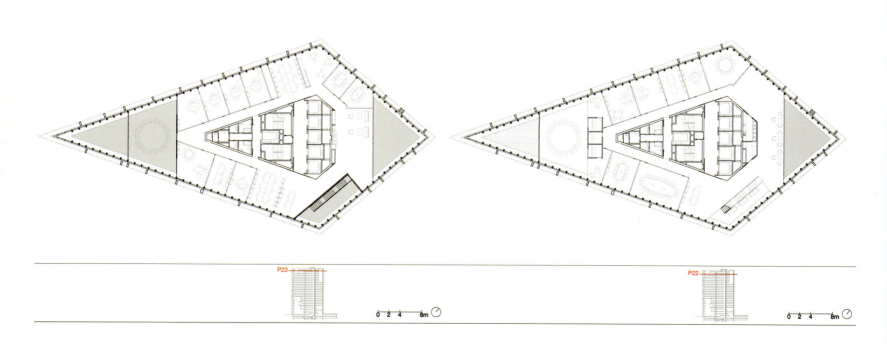

P23   0 2 4   8m

P22   0 2 4   8m

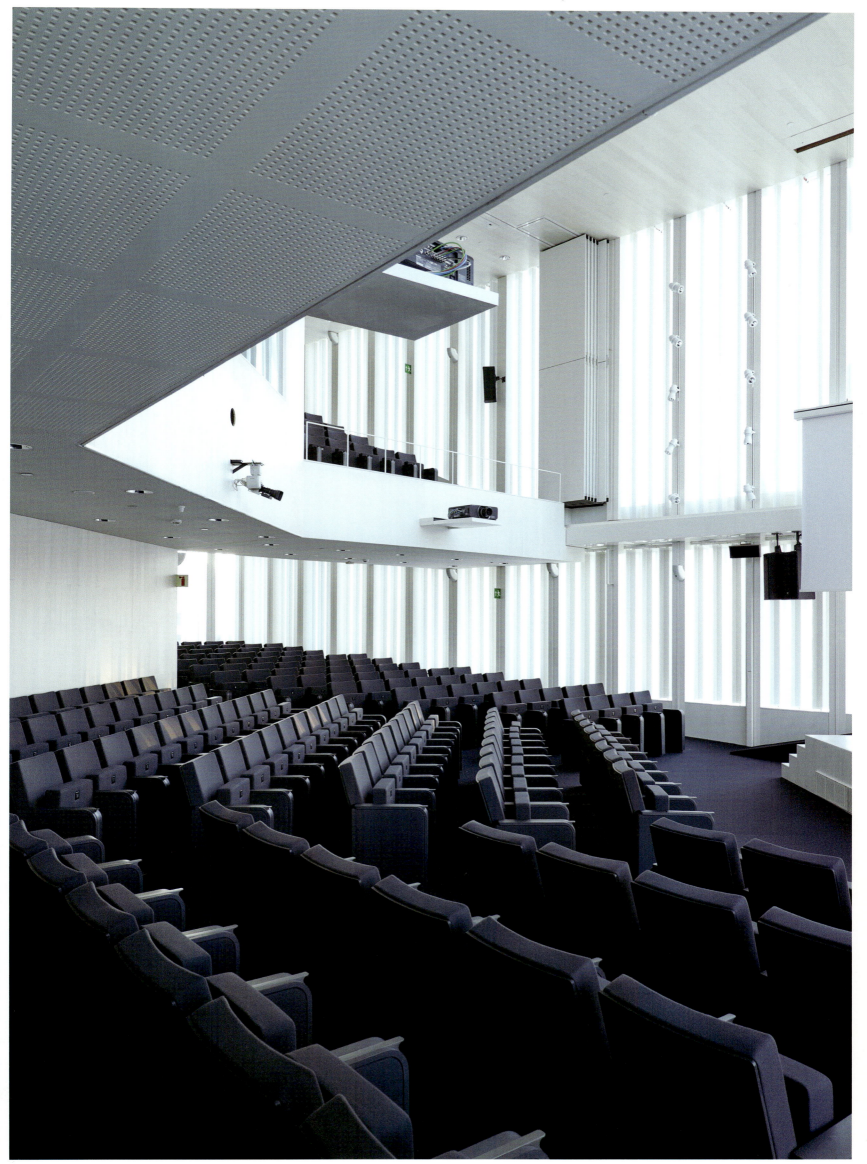

KUMHO ASIANA MAIN TOWER

锦湖韩亚主塔大楼

LOCATION: SEOUL, KOREA

ARCHITECT : SAMOO ARCHITECTS & ENGINEERS, G.S ARCHITECTS & ASSOCIATES
CLIENT : KUMHO ASIANA GROUP

AREA : 60,695 m²
PHOTOGRAPHER: YUM SEUNG HOON

The convex shape represents the taking off of Kumho Asiana Group's vision, while the concave shape reflects the history of the group as well as the assimilation of the urban context of adjacent cities. The new building expresses its new architectural image by employing these two curves.

The frontage facing Sinmunno street forms a new trend along with large corporate buildings, such as Jongno Tower, Dong-a Ilbo building, Hungkuk Life Insurance, and SK Tower. The rear side aligns with a historical axis formed by Gyeongbokgung, Gyeonghuigung and Deoksugung(palaces).

An art wall has been designed under the theme of "the land and light" in the rear, which faces Jeongdong in the absence of any high-rise buildings. Two major elements that comprise the art wall are clay art tiles representing the land, and LED scene lights representing the light. The New Kumho building will represent a milestone suggesting a new alternative through its advanced design and art wall.

Different types of glass (reflective low-E pair glass, transparent low-E pair glass) are employed to each side, in order to express different characteristic. Aluminum louver contributed to its aesthetic composition of face.

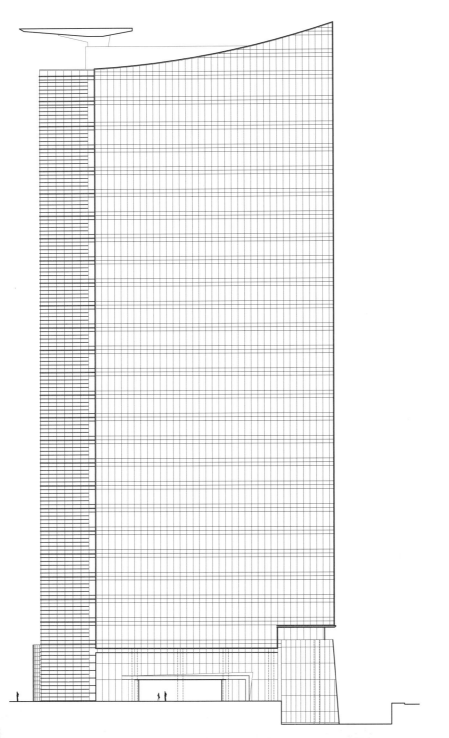
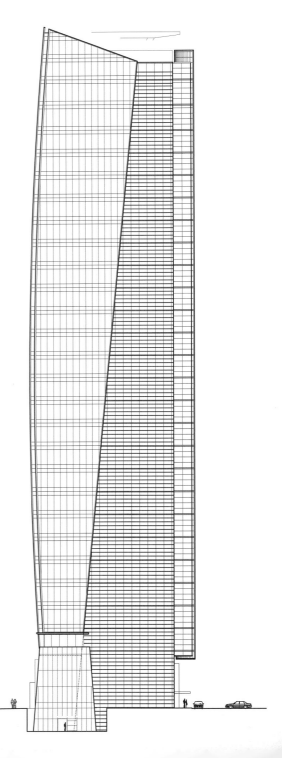

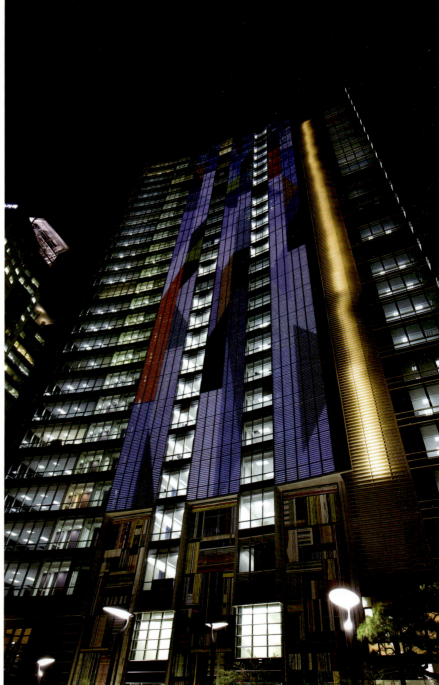

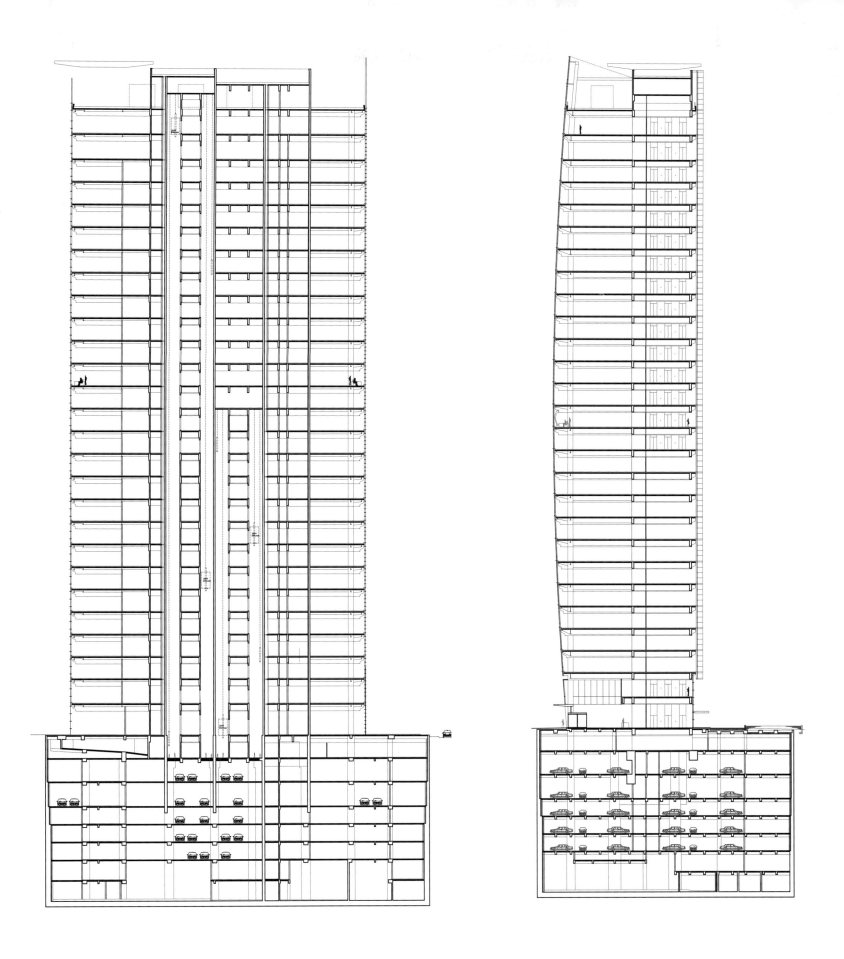

大楼凸起的形状代表着锦湖韩亚集团展翅腾飞的愿景，而凹进的形状则反映了集团的发展历史，以及对周边城市城市文脉的同化。通过采用两条曲线，新建筑呈现了新的建筑形象。

大楼正面朝向 Sinmunno 大街，与周边诸多集团大厦，如 Jongno 大厦、东亚日报大厦、Hungkuk 人寿保险大厦、SK 大厦一起，形成了一个新地标。大楼背面与景福宫、庆熙宫、德寿宫（宫殿）形成的历史轴线对齐。

在大楼后面设计了一个艺术墙，艺术墙以"地与光"为主题，在没有其他高层建筑的情况下，艺术墙和 Jeongdong 正好正面对面。艺术墙由两个主要元素组成，即代表地的黏土艺术瓷砖和代表光的 LED 场景灯。新的锦湖韩亚主塔大楼，通过其先进的设计理念和艺术墙，将成为又一个建筑设计的里程碑。

大楼每一侧都使用了不同类型的玻璃（反射型低辐射双层玻璃、透明型低辐射双层玻璃），以表达各自不同的特色。设计的铝百叶窗也有助于提升大楼的美观度。

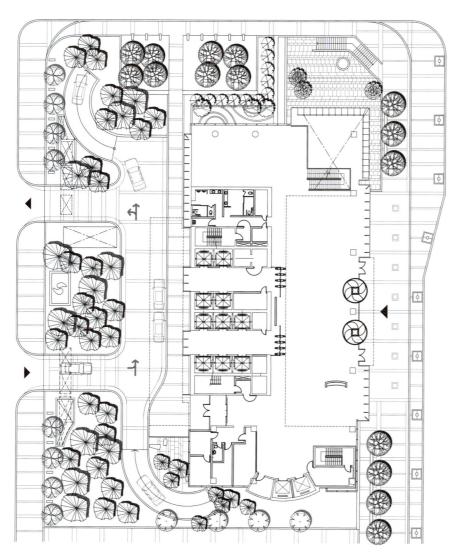

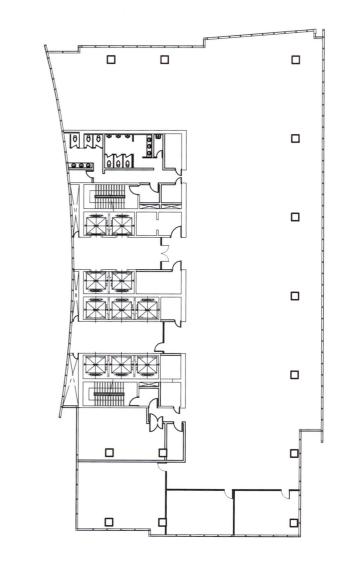

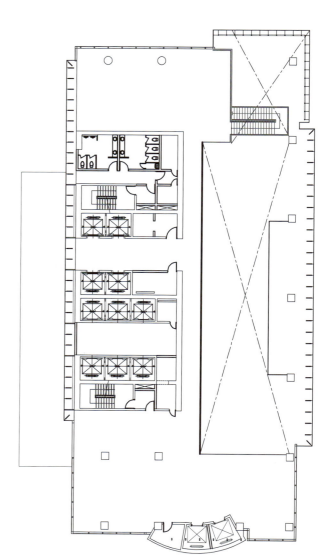

BREEZÉ TOWER, OSAKA

大阪和风大厦

LOCATION:
OSAKA, JAPAN

ARCHITECT: INGENHOVEN ARCHITECTS

AREA: 76,000 m²

The Sankei Building Corporation, the client for this 177-meter tall high-rise building, is a subsidiary of the Fuji Broadcasting Company of Japan. The building on National Road No. 2 in the Umeda business district of Osaka replaces a building from the 1950s. The floor plan allows for a highly flexible usage. A nine-storey podium, with a large concert and event hall for 960 people, restaurants and conference rooms, along with a shopping mall, creates an important new pedestrian connection in Umeda.

The Breezé Tower is the first environmentally-friendly skyscraper in Japan with a double-skin glass facade that allows for natural ventilation of the interiors. Qualified as "S-Class" according to the Japanese CASBEE system, the Breezé Tower has received the highest possible rating for ecological architecture in Japan.

这座 177 米高的大厦属于日本富士电视公司的子公司三惠建筑公司。大厦位于大阪梅田商贸区国道 2 号，取代了一座建于 20 世纪 50 年代的建筑。

建筑平面考虑到了多样性的用途。一个九层高的裙楼，内设可容纳 960 人的大型音乐厅和会议活动场所、餐馆、会议室，以及购物中心，成为了梅田地区的一个新的重要的行人活动区。

和风大厦采用双层玻璃外立面，让内部得到自然通风，是日本第一座环境友好型摩天大厦。和风大厦达到了日本 CASBEE 建筑物综合环境性能评价体系的 S 级（特优）认证，成为了日本评价最高的生态环保型建筑。

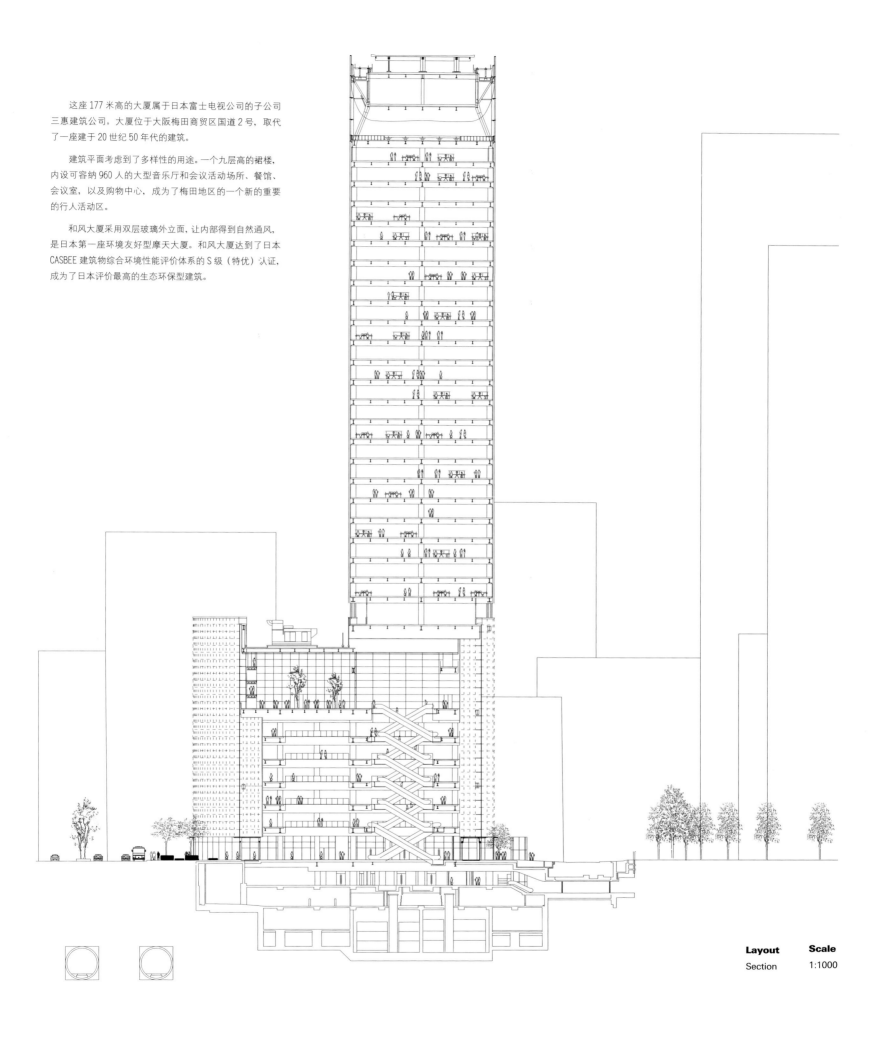

Layout   **Scale**

Section   1:1000

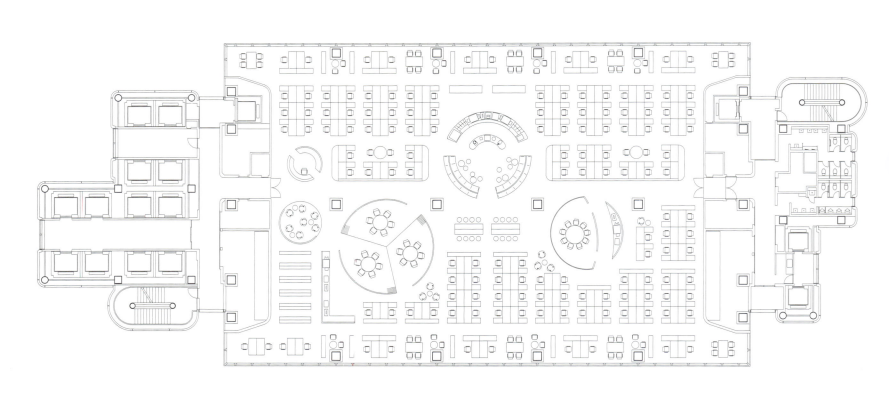

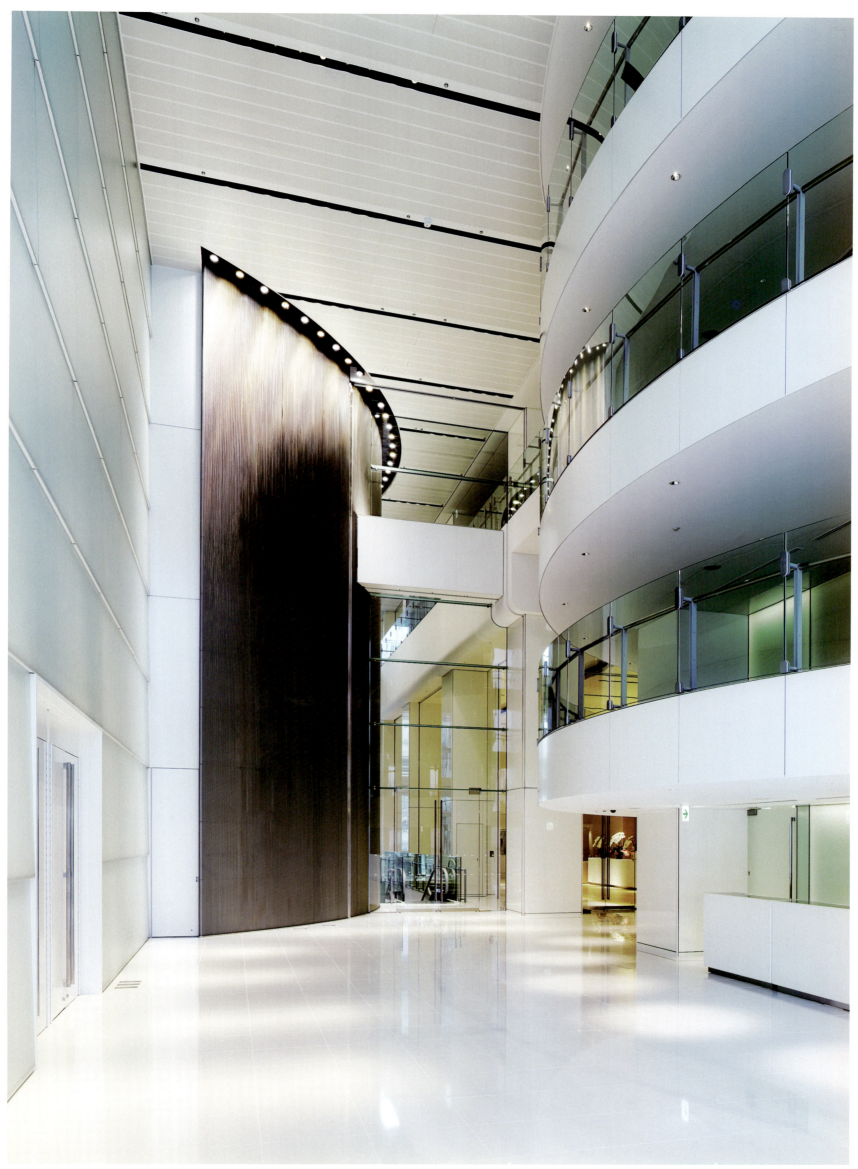

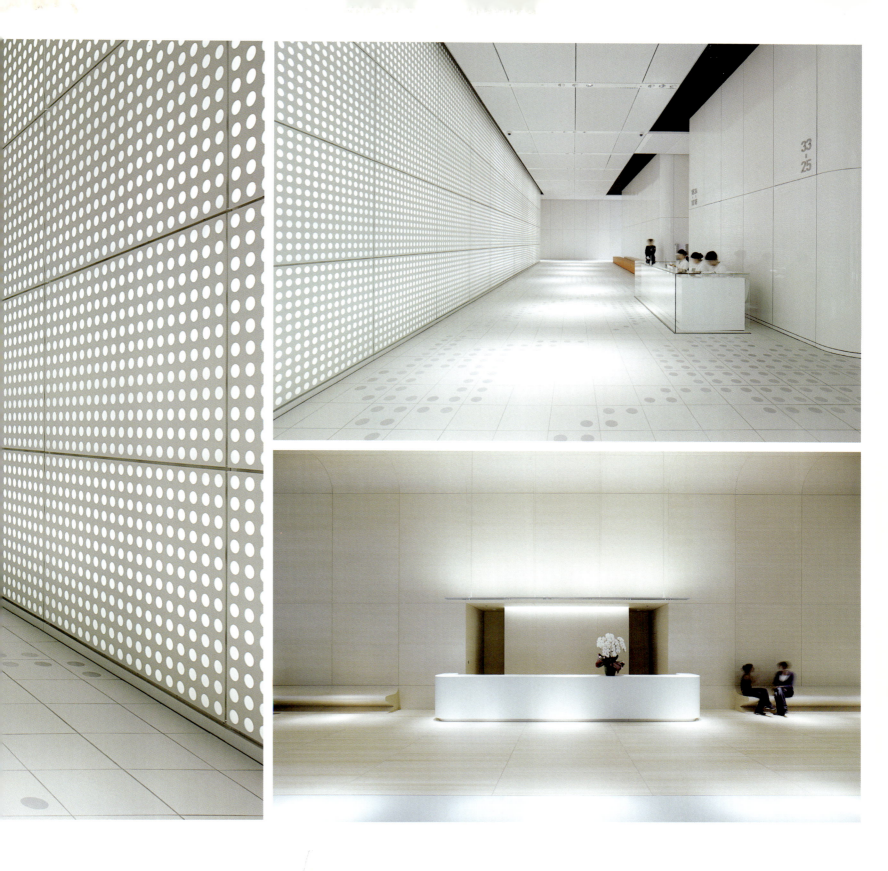

VIENNA DC TOWERS

维也纳DC双塔

LOCATION:
VIENNA, AUSTRIA

ARCHITECT: DOMINIQUE PERRAULT ARCHITECTURE
CLIENT: WED (WIENER ENTWICKLUNGSGESELLSCHAFT FÜR DEN DONAURAUM AG)
TOTAL BUILT AREA: 229,000 m²

ASSOCIATED ARCHITECT: HOFFMANN & JANZ ARCHITECTES, VIENNA, AUSTRIA
RENDERING: DOMINIQUE PERRAULT /BEYER.CO.AT / DPA / ADAGP

The design of the two high-rise towers for the Donau-City in Vienna represents the concluding phase of a development extending over several decades: on what was originally a municipal rubbish tip the UNO-City was erected (1973~1979), tentative plans to hold the 1995 Vienna-Budapest EXPO here were soon abandoned, as a result architects Krischanitz and Neumann (commissioned by WED AG) produced an urban design masterplan for the area in 1992. The outcome is an entirely new urban district with a diverse range of functions. With a total area of some 17.4 hectares and total investment of roughly € 2 billion, the VIENNA DC Donau-City is by far Austria's largest real estate development.

A total of approximately 1.7 million cubic meters will be built, which equates to about 500,000 m² of gross space. Not quite two-thirds of the buildings have been completed and let. The international competition that followed in 2002 for the design of the remaining undeveloped third of the Donau-City was won by Dominique Perrault. To ensure that the development would provide the entire Donau-City site with a new kind of urban quality, Perrault's urban planning guideline project employs a number of different design measures: firstly his project continues the elevated slab of the Donau-City to the banks of the Neue Donau in the form of a generously dimensioned terrace providing direct access to the river. Secondly rather than interpreting the two high-rise towers as independent buildings Perrault treats them as the corresponding halves of a block that open towards the city and the Neue Donau with a space-defining gesture.

维也纳多瑙城两个高层大厦的设计，代表了一个延续了几十年的开发项目的收尾阶段：UNO城的选址最初是一个城市垃圾场（1973~1979），1995年维也纳－布达佩斯世博会原本设想在这里举行，但该计划很快就搁浅了。于是，建筑师Krischanitz和Neumann（受WED AG委托）在1992年为该地区制定了一个城市设计总体规划，目标是建造一个全新的、能够满足不同功能需求的城市社区。占地总面积约17.4公顷，总投资约2亿欧元，维也纳DC多瑙城是迄今为止奥地利最大的房地产开发项目。

共约170万立方米的建筑将被建设，这相当于约50万平方米的总建筑面积。不到三分之二的建筑已经完成并出租出去。2002年，Dominique Perrault在国际竞争中赢得多瑙城剩余三分之一建筑的设计权。为了确保开发能给整个多瑙城一种新的城市品质，Perrault的城市规划采用了一些不同的设计措施：首先，项目继续采用了多瑙城通往新多瑙河岸边使用的高架板，形成与河边相连的大尺寸的台阶；其次，Perrault没有把两幢高楼视为孤立的建筑，而是看作两个相对的半块，以特定的空间姿势朝向城市空间和新多瑙河。

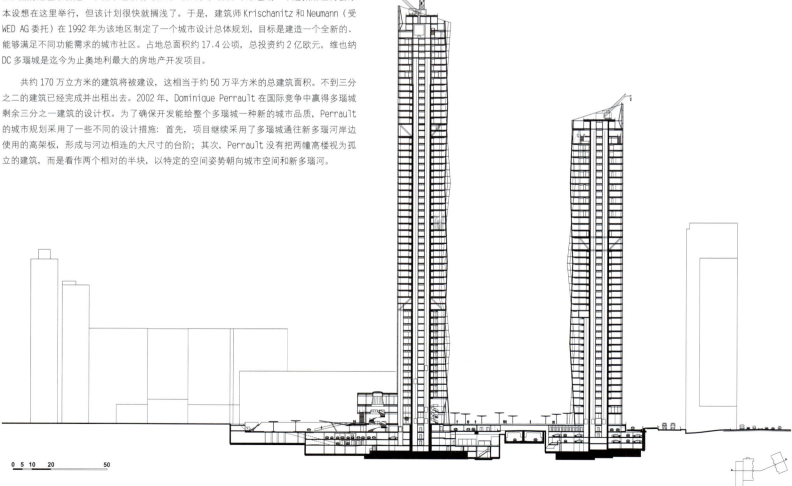

0 5 10   20   50

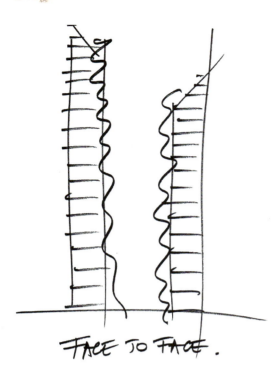

FACE TO FACE.

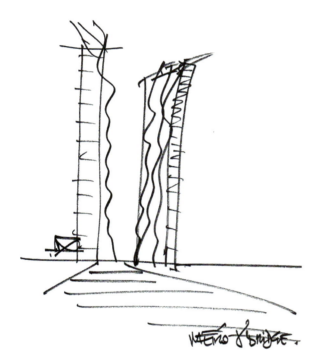

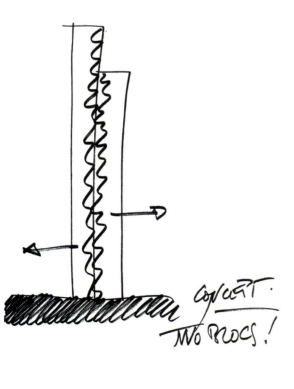

CONCEPT.
TWO BLOCS !

ONE RAFFLES PLACE TOWER 2 莱佛士坊 2 号大厦

LOCATION: SINGAPORE

ARCHITECT: TANGE ASSOCIATES
ASSOCIATE FIRM: SAA ARCHITECTS PTE. LTD.
MEP ENGINEER: BECA CARTER HOLLINGS & FERNER PTE. LTD.
CONTRACTOR: SATO KOGYO - HITACHI PLANT JOINT VENTURE

CLIENT: OUB CENTRE LIMITED
AREA: 43,718.80 m²
PHOTOGRAPHER: MARC TEY

One Raffles Place Tower 2, located in the Central Business District (CBD) of Singapore, has taken its place next to the iconic Tower 1 designed by Kenzo Tange, the site formerly known as OUB Center. Tower 1, built in 1986, was the tallest building in the world outside of the US at the time, and still one of the tallest in Singapore.

Designed to stand alongside Tower 1, the new Tower 2 is a contemporary building that respects the design elements of Tower 1. Tower 1 is sleek but powerful with its solid aluminum facade. Tower 2's elegant glass curtain wall maintains the sleek image of Tower 1 and also expresses its own character as a new addition to the landmark complex.

The triangular motif of Tower 1 is incorporated into the design of Tower 2 from ground level to rooftop. Never-seen before sculptural illumination adds beauty to the night view of the CBD area. The architect/artist collaborative installation of a waterfall, a drawing of Japanese painter, Hiroshi Senju, consists of porcelain panels with the picture of a waterfall with real water flowing over the artwork. It is located in the center of main entrance, and creates a refreshing, peaceful atmosphere in the lobby in contrast to the heat and bustle of the CBD.

Not simply its appearance, but One Raffles Place Tower 2 is credited with its functional aspect, by receiving the Singapore Building and Construction Authority (BCA) Green Mark Platinum Award and being nominated for the International High-rise Award, which is bestowed on a building that stands out for its special aesthetics, pioneering design, integration into its urban context, sustainability, innovative technology and cost effectiveness. As a new landmark, this new tower will inject new enthusiasm into the majestic Singapore City Skyline.

位于新加坡中央商务区的莱佛士坊2号大厦，就在Kenzo Tango设计的1号楼旁边，这个位置以前叫海外联合银行中心。1号楼建于1986年，当时是世界上除了美国建筑之外的最高的建筑，即便在现在，也是新加坡最高的建筑之一。

2号大厦就在1号楼的旁边，虽然是一栋当代建筑，但在设计上仍然沿袭了1号楼的一些元素。1号楼坚固的铝板外立面极富光泽，充满气势。2号大厦采用了玻璃幕墙，既传承了1号楼的光泽和雅致，又体现了一栋新地标建筑的独特个性。

1号楼的三角形设计主题也融入了2号大厦的设计中，从底层到顶层，无处不在。前所未有的雕塑夜景照明设计，为中央商务区的夜景增添了色彩。建筑设计师与艺术家的通力合作，让日本艺术家千住博的绘画作品《瀑布》印在了陶瓷面板上，并有真的水流从表面一流而下。这幅作品就在主入口的中心，在大厅营造出宁静、清新的氛围，与整个中央商务区的喧嚣与繁华形成鲜明的反差。

莱佛士坊2号大厦外观不俗，功能更优。它曾荣获新加坡建设局"绿色标志"白金奖，以及国际高层建筑奖的提名。这些荣誉都在表明，这座建筑具有独特的美学价值、创新的设计，以及与都市环境融为一体的可持续性、创新技术和成本控制方案。作为一座新的地标，2号大厦将为新加坡宏伟的城市风貌增添新的活力。

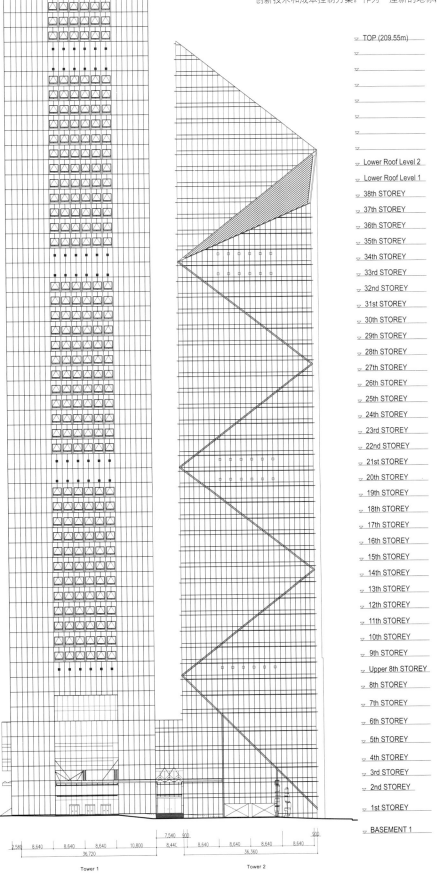

TOP (209.55m)

Lower Roof Level 2
Lower Roof Level 1
38th STOREY
37th STOREY
36th STOREY
35th STOREY
34th STOREY
33rd STOREY
32nd STOREY
31st STOREY
30th STOREY
29th STOREY
28th STOREY
27th STOREY
26th STOREY
25th STOREY
24th STOREY
23rd STOREY
22nd STOREY
21st STOREY
20th STOREY
19th STOREY
18th STOREY
17th STOREY
16th STOREY
15th STOREY
14th STOREY
13th STOREY
12th STOREY
11th STOREY
10th STOREY
9th STOREY
Upper 8th STOREY
8th STOREY
7th STOREY
6th STOREY
5th STOREY
4th STOREY
3rd STOREY
2nd STOREY
1st STOREY
BASEMENT 1

NORTH ELEVATION SCALE 1:750 (A3)
0   5   10   15   20M

2,580 8,640 8,640 8,640 10,800
36,720
Tower 1

7,540 900 8,440 8,640 8,640 8,640
36,360
Tower 2

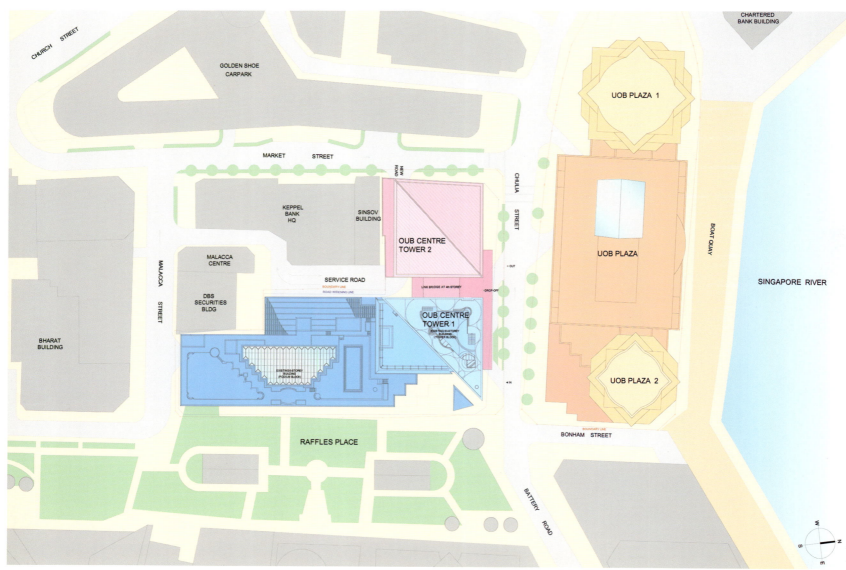

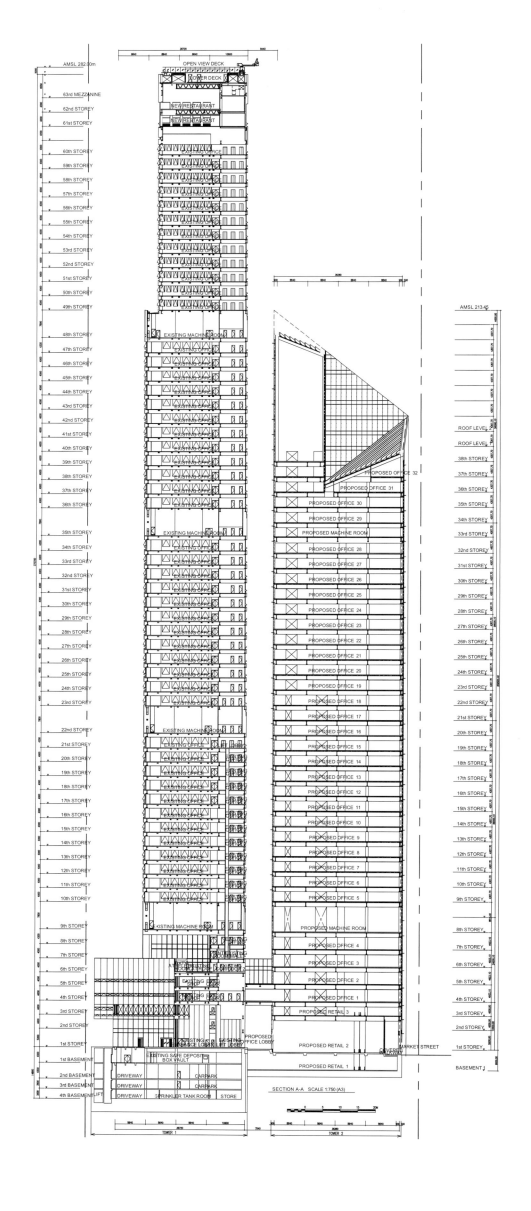

<u>1ST STOREY PLAN (TOWER 2)</u>

SCALE 1:750 (A4)

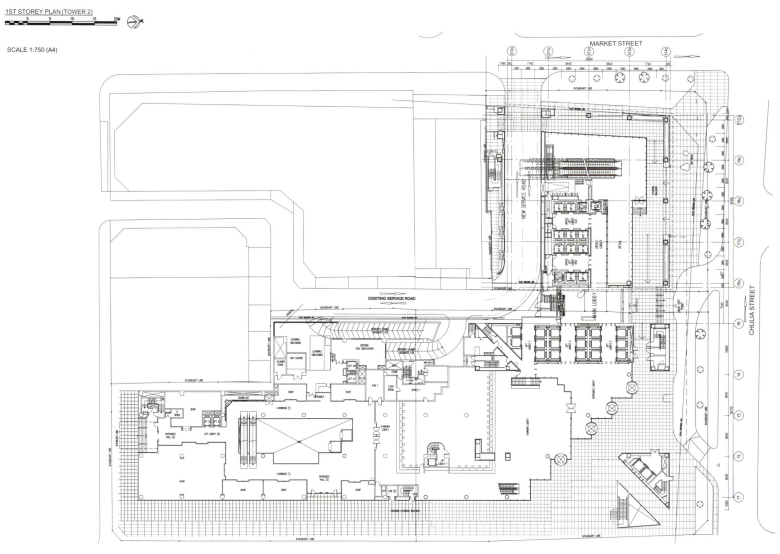

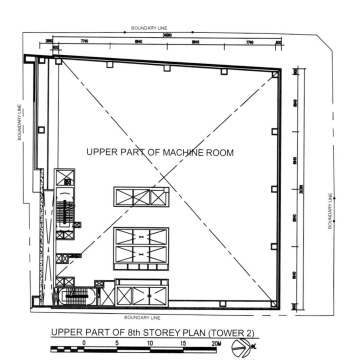

UPPER PART OF 8th STOREY PLAN (TOWER 2)

0 5 10 15 20M

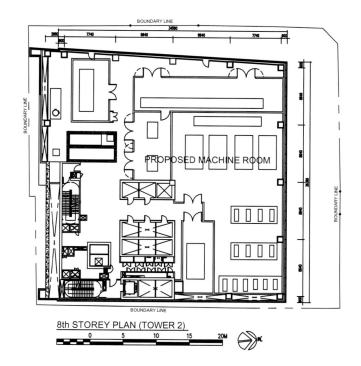

8th STOREY PLAN (TOWER 2)

0 5 10 15 20M

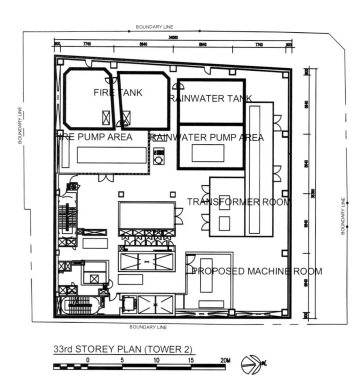

33rd STOREY PLAN (TOWER 2)

0 5 10 15 20M

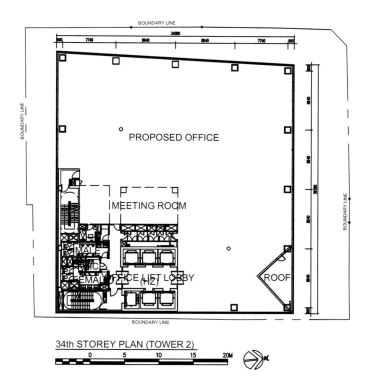

34th STOREY PLAN (TOWER 2)

0 5 10 15 20M

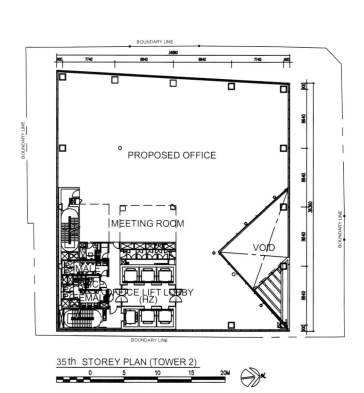

35th STOREY PLAN (TOWER 2)

0 5 10 15 20M

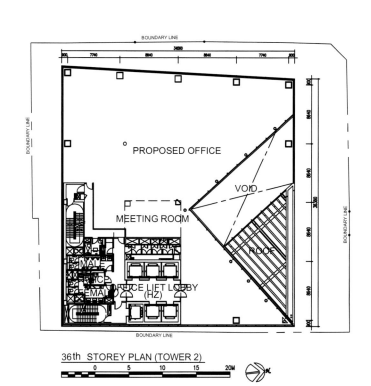

36th STOREY PLAN (TOWER 2)

0 5 10 15 20M

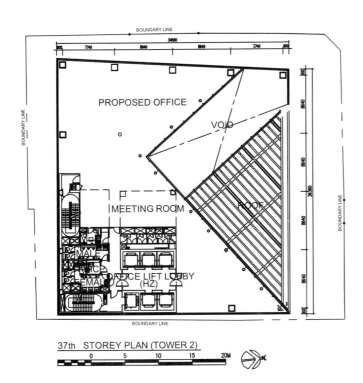

37th STOREY PLAN (TOWER 2)

0 5 10 15 20M

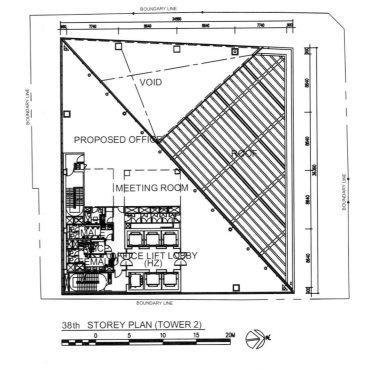

38th STOREY PLAN (TOWER 2)

0 5 10 15 20M

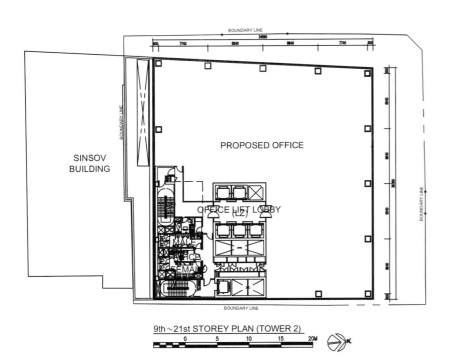

9th~21st STOREY PLAN (TOWER 2)

0 5 10 15 20M

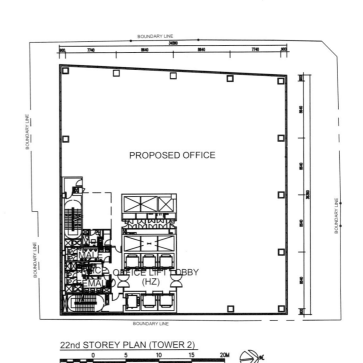

22nd STOREY PLAN (TOWER 2)

0 5 10 15 20M

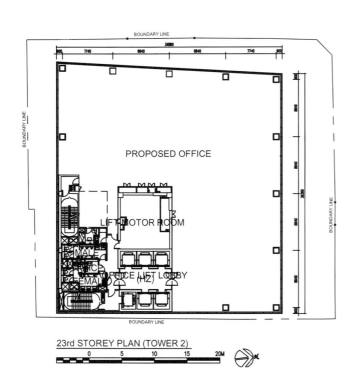

23rd STOREY PLAN (TOWER 2)

0 5 10 15 20M

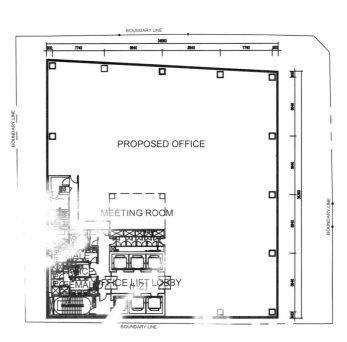

24th to 3?rd STOREY PLAN (TOWER 2)

0 5 10 15 20M

SHENZHEN NEO TOWER

深圳绿景大厦

LOCATION:
SHENZHEN, CHINA

ARCHITECT: CCDI
AREA: 8,700 M² (SITE), 104,200 M² (GROSS FLOOR)

In the visual arts, contrast is what gives a shape definition and meaning. With this in mind, the NEO Tower is designed as a modern landmark for the city of Shenzhen using such ideas. Built as a conversation between bold lines and sharp curves, the tower suggests the breakthrough technology and commercial prowess through contrasting colors and shape. With its sharp edges, a play of convex and concave lines distinguish each facade; while one side curves inward, the opposite side curves outward; dark glass and steel on one side is balanced by light materials on the other. The curved facades of the building lay in direct contrast to the rectangular side facades. Two small podiums reinforce this positive/negative relationship, taking on the design and color of their larger compatriots.

The convex shape of the main facade also plays a role in the urban planning to relax the tight relation and distance with the nearby Zhengxie Tower. Contextually, the NEO Tower is lifted on the east corner of the site, leaving space for a city plaza. This opens up the area around the building, creating an active and multi-purpose space that connects the site to the city axis.

The podiums, angled towards the sky, echo the principle shape of the main towers. Large LED screens allow for images and text to appear on the podium skin. The roof area is a complex play of angles which complement each other through a triangular structure, like dueling forces of light and dark. In addition to its aesthetic design, the building uses Low-E hollow glass to conserve energy. A contrast between curved/flat shapes, horizontal/vertical lines, and dark/light color, forms this active and bold design.

South Elevation

East Elevation

North elevation

West elevation

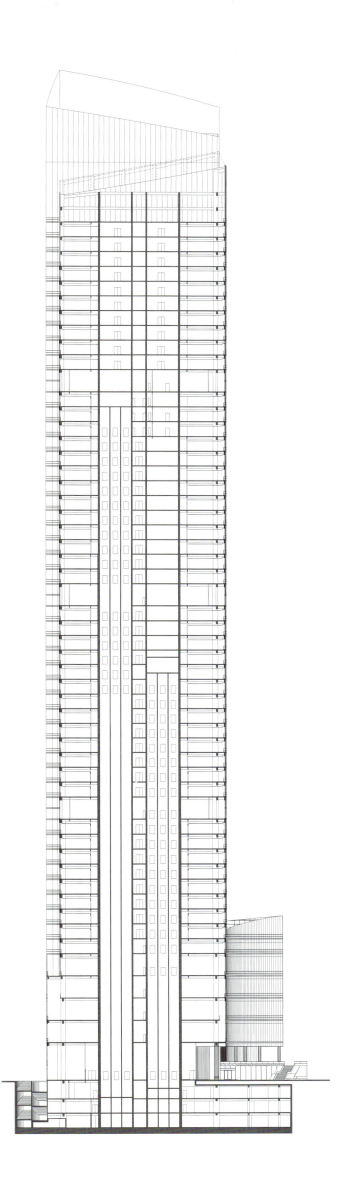

在视觉艺术中，对比赋予形状具体的定义和意义。绿景大厦正是运用这个理念来设计，从而成为深圳的现代地标建筑。在粗线条和锐曲线之间的转换中，绿景大厦通过对比鲜明的色彩和形状显示出突破性技术和超凡的商业实力。凹凸的线条通过其锐利的边缘使每个立面都尤为突出；当一边凹进去时，另一边则凸起来；一边的黑色玻璃和钢铁与另一边的浅色材料达到平衡。大厦的弯曲立面和矩形立面形成鲜明的对比。两个小型裙房同样运用了大裙房的设计和色彩，从而更加凸显出这种正／反关系。

主立面的凸出形状同样在城镇规划方面缓和了其和旁边的政协大厦的关系，并扩大了它们之间的距离。因此，绿景大厦位于该地块的东面，并为一个城市公共广场预留了空间。这个设计开拓了建筑周围的区域，创造出一个连接该地块和城市中轴线的、积极的多功能的空间。

斜向朝天的裙房和主楼的主要形状一样。裙房的表面上的大型 LED 屏幕用以投放文字和图片。屋顶由不同的角组成，这些角因其三角形结构互相补充，就像光和暗的较量。除了它的美学设计外，绿景大厦还运用低辐射中空玻璃来节约能源。弯曲和扁平的形状、水平和垂直的线条、深浅颜色的对比形成了这个积极而大胆的设计。

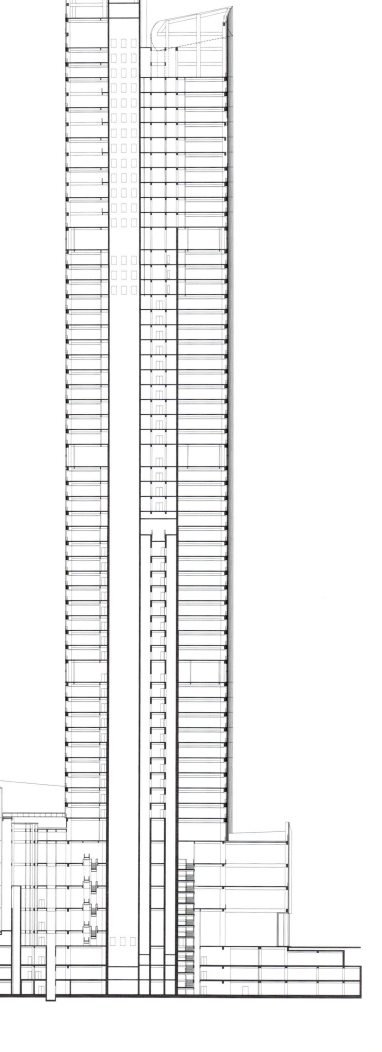

ARRAYA TOWER

Arraya 大厦

LOCATION:
KUWAIT CITY, KUWAIT

ARCHITECT: FENTRESS ARCHITECTS
ASSOCIATE ARCHITECT: PAN ARAB CONSULTING ENGINEERS (PACE)

OWNER: SALHIA REAL ESTATE COMPANY
PHOTOGRAPHER: NICK MERRICK, PAWEL SULIMA

Arraya Tower is one of Kuwait's tallest skyscrapers and holds the distinction of being the World's 4th Tallest Building completed in 2009. Fentress discovered a way for the Arraya Tower to both make peace with, yet stand out from, its neighbors by weaving complementary design schemes and materials together to celebrate the Tower's geographic and cultural context.

The Tower's height is awe-inspiring, yet the building's mass presents a slender profile.

Three distinct zones and varied single-tenant floor plates break down its mass. Limestone and metal cladding adorn a glass curtainwall, and a vocabulary of punched windows harmoniously unifies the building with its neighbors. The design integrates state-of-the-art communications technology with the latest in high rise requirements. It offers flexibility: raised floors allow easy access to power and teledata cabling, while tenant space can be easily reconfigured.

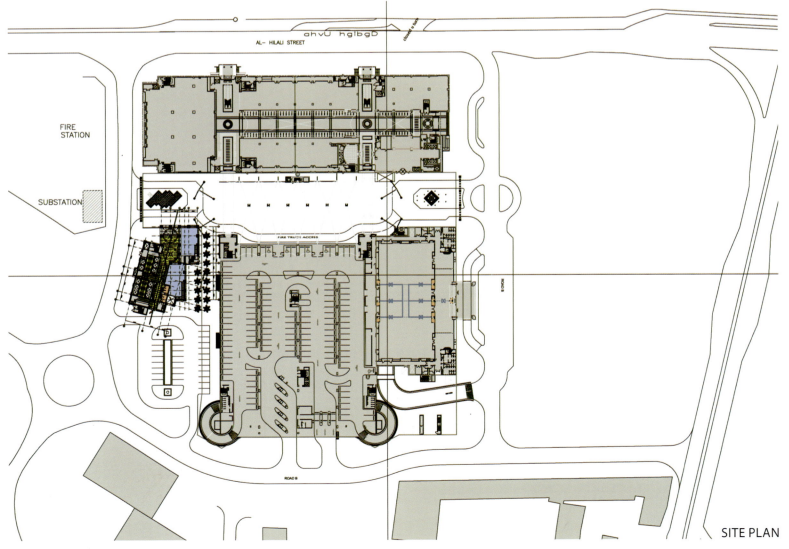

SITE PLAN

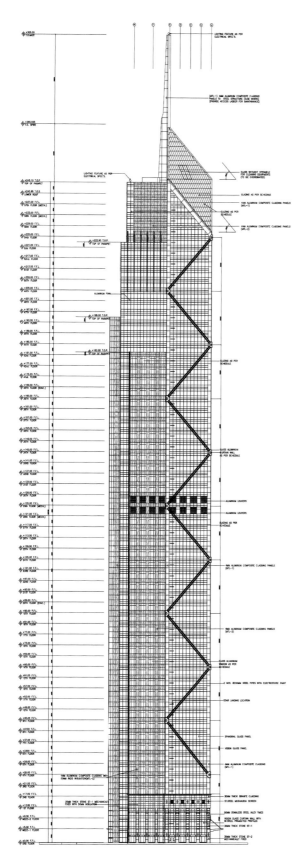

EAST ELEVATION

WEST ELEVATION

Arraya 大厦是科威特最高的摩天大楼之一，享有 2009 年建成的世界上第四高楼的美誉。为了突出大厦的地理和文化特色，Fentress 通过采用互补的设计方案和建筑材料，找到了一个既让 Arraya 大厦融入周边环境，又与众不同的方法。

大厦的高度令人惊叹，建筑主体却呈现出纤细的轮廓感。三个不同的区域和多样的单租户楼面，把大楼主体分成了几个组成部分。石灰墙和金属板装饰了玻璃幕墙、打孔窗户，让大厦与周围的建筑和谐统一。该设计集成了高层建筑要求的最新、最先进的通信技术。架高地板方便接入电源和远程数据电缆，承租人空间可以很容易地重新配置，这些都为建筑提供了灵活性。

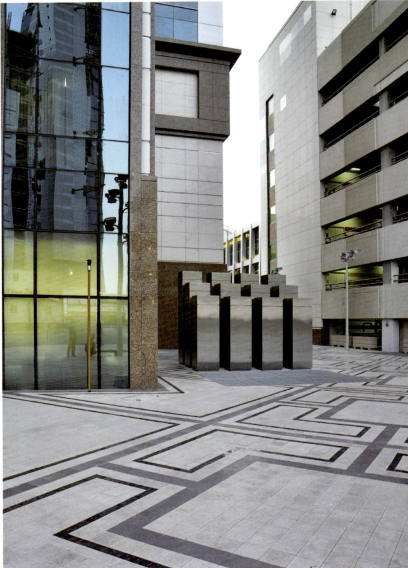

CHENGDU MINGYU FINANCIAL PLAZA

成都明宇金融广场

LOCATION:
CHENGDU, SICHUAN
PROVINCE, CHINA

ARCHITECT: MING LAI ARCHITECTS INC.
AREA: 119, 183 m²

The inspirational idea of this project comes from the relic discovered in San Xing Dui, Chengdu – the Yu Zhang or known as the Jade ornament. Combining the relic culture and shape of the jade ornament, designing of the building uses a lot of the arcs similar to the jade ornament. Standing at 200 meters tall, the building stands comfortably without the hindrance of a podium below. Surrounded by square office blocks, height of the building combine with the arcs and curves makes the building standout as a magnificent landmark.

The tower comprises of offices, hotels and amenities. The building is divided into 4 main functions namely 5 floors of basements, 1~7 floors of lobbies, meeting and conference rooms and restaurants, 8~31 floors of low and high rise office spaces and 33~47 floors of five star hotel rooms. Oval shape planning positioned diagonally on the site provides an unobstructed 180 degrees view. Main tower block together with the New Century building located on the west side creates a harmonious dialogue between the two buildings. Angled towards the south is the main street Dong Da Jie, East Avenue which enables the main facade facing the busy downtown demonstrating a welcoming gesture.

Facade design uses the notion of different planes on the glass curtain walls which create a light and shadow effects enhancing the richness in the facade elements. At the same time, using the glass curtain-walls and steel structure arc shape of the building to strengthen the idea and shape of the jade ornament. During the night, lighting effects will be use to highlight the unique form of the building.

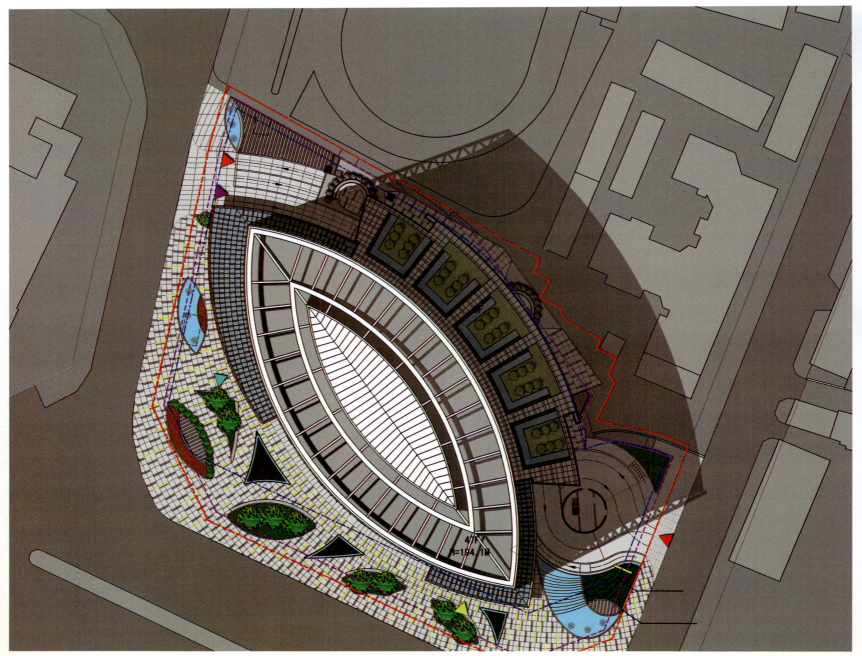

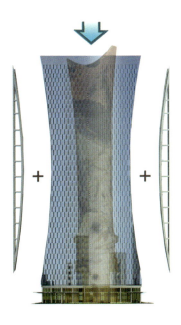

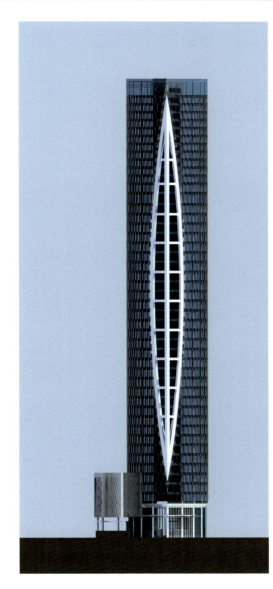

Jade cicada was a very important sacrificial vessel in ancient China. There is a saying from "Zhou Li (The Rites)" that jade wall, jade slice, jade tablet, jade cicada, jade pendant and tiger-shape jade are six kinds of jade ritual objects to treat everybody courteously, and jade cicada is to set up brigade and to help the soldiers to defend enemies. Jade cicada is not only a ritual object for the ancients to worship Heaven, to salute the Sun and to pray for the good harvest, but also a symbol of power and dignity. Sanxingdui Ruins in Sichuan Province has the most jade cicada in China.

玉璋在中国古代是一种极为重要的礼器，《周礼》有"玉作六器（璧、琮、圭、璋、璜、琥）以礼天地四方"，"牙璋……以起军旅，以治兵守"的记载，玉璋既是古人祭天、拜日、祈年等的礼器，又是权利和尊贵社会地位的象征。四川三星堆遗址是目前全国出土玉璋最多的地区。

DESIGN GOAL
建筑设计意向

DESIGN STUDY
设计过程

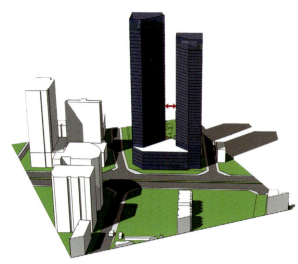

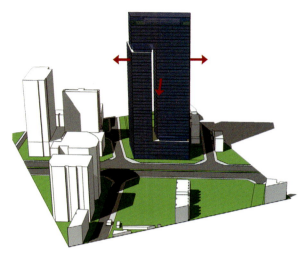

Separate Twin Tower

1. Functions of hotel and office are vertically separated.

2. Building density increases, landscaping inside the site decreases, sight between the towers interrupts each other.

3. The rate between the core tube area and used area is high, the economy is poor.

4. Building height is comparably lower, less landmark.

独立双塔

1. 酒店与办公的垂直功能分别独立。

2. 建筑密度增加，减少基地内绿化，塔楼之间视线相互干扰。

3. 核心筒面积与使用面积的比例高，经济性差。

4. 建筑高度偏低，地标性差。

Overlapped Twin Towers

1. Functions of hotel and office are vertically separated.

2. Only 3 sides of the building will get natural daylight.

3. Oppression against on the east street in the urban space is strong.

4. The rate between the core tube area and used area is high, the economy is poor.

5. Part of sight is interrupted.

双塔叠合

1. 酒店与办公的垂直功能分别独立。

2. 酒店与办公体量间仅有三面采光。

3. 城市空间上对东大街的压迫感强。

4. 核心筒面积与使用面积的比例高，经济性差。

5. 局部视线干扰。

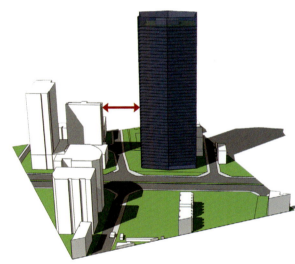

Single Arc-shaped Tower

1. Functions of hotel and offices are disposed centrally. Public equal share is little, the economy is high.

2. Sights between the hotel and offices won't interrupt each other.

3. The curve forms a front square and creates urban spaces.

4. The vision is broad. The main vision is directed to the main landscape axis of the city.

5. An unsealed space is formed with the west-side Century Plaza in a welcoming spirit.

6. The dynamic and curved building stands out of the neighboring square-shaped buildings, to become an obvious landmark.

7. The building won´t block the daylight from surrounding buildings.

单体弧形

1. 酒店与办公的垂直功能集中配置，公摊系数小，经济性强。

2. 酒店与办公视线互相干扰。

3. 旋转形成前区广场，创造都市空间。

4. 视野开阔，主要视线方向迎合城市主要景观轴线。

5. 与用地西侧的新世纪广场形成半围合的建筑空间，体现欢迎的姿态。

6. 动态弧线易从周边方正的建筑群落中脱颖而出，形成明显的地标性建筑。

7. 对周边建筑无日照遮挡。

Single Rectangle Tower

1. Functions of hotel and offices are disposed centrally. Public equal share is little, the economy is high.

2. Sights between the hotel and offices won't interrupt each other.

3. Building form is rigid and too huge. Concordance with the urban context is lack of creativity.

4. Oppression against on the east street in the urban space is strong.

5. The building will block the daylight from surrounding buildings.

单体矩形

1. 酒店与办公的垂直配置，公摊系数小，经济性强。

2. 酒店与办公视线无互相干扰。

3. 建筑体量僵硬，过于庞大，在城市空间和谐性上，缺乏新意。

4. 城市空间上对东大街的压迫感较强。

5. 对周边建筑日照遮挡严重。

MASSING STUDY

整体建筑意象的灵感来源于成都三星堆的出土文物——玉璋。结合玉璋的文化延伸及体态暗示，设计在平面和立面造型上大量运用弧线，200米的高度形成舒展的体态，塔楼不受裙楼阻隔，高度和弧线的完美结合使建筑在本地区大量的方正建筑中脱颖而出，并演变成气势恢宏的地标性建筑。

塔楼将办公、酒店及其他配套功能垂直结合在一个体量当中。由5层的地下室，1~7楼的入口大厅、会议室及餐饮区，8~31楼的低区及高区甲级办公区，33~47层的五星级酒店等

4个部分组成。平面布局上，橄榄型平面对角线布置于矩形基地内，平面造型本身具有完整的180度观景视角。主体塔楼结合西边的新世纪大楼形成和谐的半围合建筑形态，与南边的东大街形成一定的夹角，使主要的立面朝向市中心繁华的都市景观，展现出一种欢迎的姿态。

外观设计上，通过玻璃幕墙单元间进退而产生的光影效果来丰富外立面的层次，同时利用贯穿塔身的弓形弧线框架造型结构与玻璃幕墙，加强了塔身的玉璋造型，并且，夜间照明也被运用到了弓形上，更加彰显了建筑的独特性。

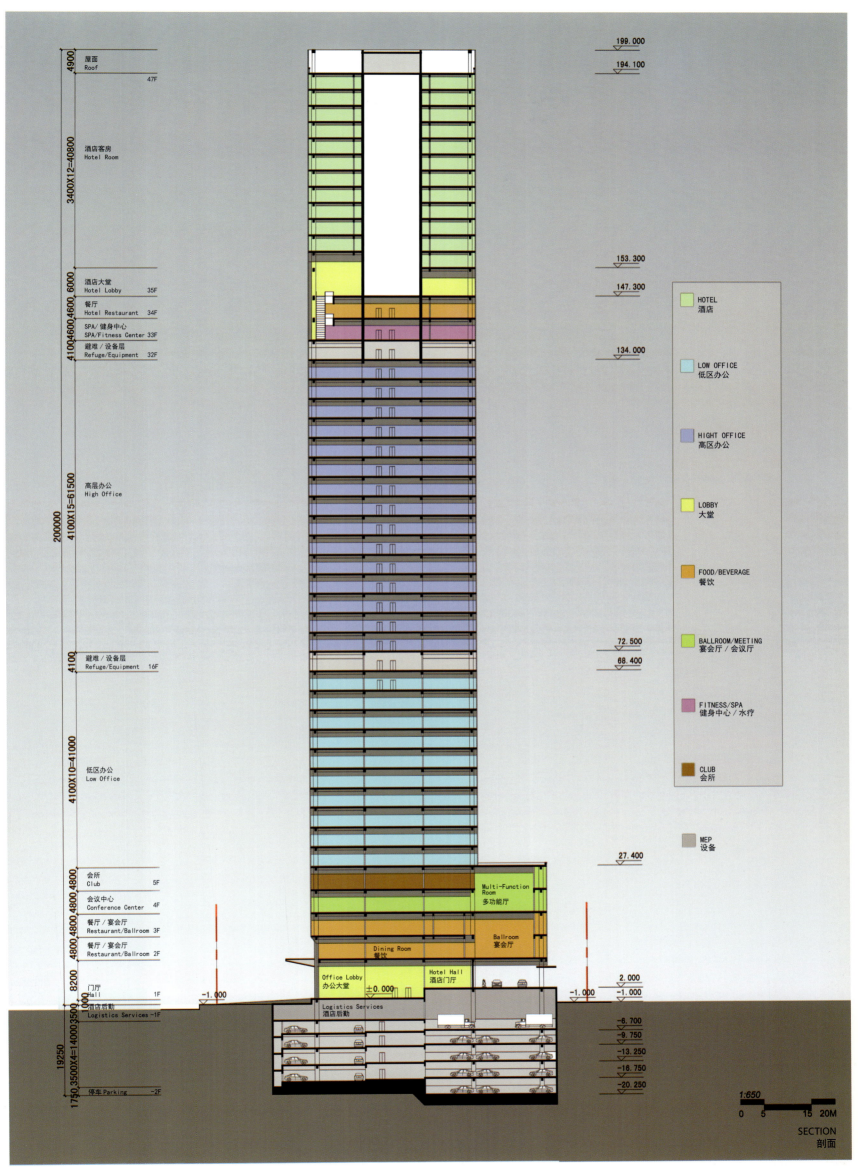

屋面
Roof   47F   199.000
194.100

酒店客房
Hotel Room

3400X12=40800

153.300
酒店大堂
Hotel Lobby   35F   147.300
餐厅
Hotel Restaurant 34F
SPA/健身中心
SPA/Fitness Center 33F
避难/设备层
Refuge/Equipment 32F   134.000

高层办公
High Office

4100X15=61500

200000

72.500
避难/设备层
Refuge/Equipment 16F   68.400

低区办公
Low Office

4100X10=41000

27.400
会所
Club   5F
Multi-Function Room
多功能厅
会议中心
Conference Center 4F
餐厅/宴会厅
Restaurant/Ballroom 3F
Ballroom
宴会厅
餐厅/宴会厅
Restaurant/Ballroom 2F
Dining Room
餐饮
门厅
Hall   1F   Office Lobby   Hotel Hall   2.000
办公大堂   酒店门厅
±0.000   -1.000   -1.000
-1.000
酒店后勤
Logistics Services -1F   Logistics Services
酒店后勤   -6.700
-9.750
-13.250
-16.750
停车 Parking   -2F   -20.250

19250 3500X4=14000 3500 1750

HOTEL
酒店

LOW OFFICE
低区办公

HIGHT OFFICE
高区办公

LOBBY
大堂

FOOD/BEVERAGE
餐饮

BALLROOM/MEETING
宴会厅/会议厅

FITNESS/SPA
健身中心/水疗

CLUB
会所

MEP
设备

1:650
0   5   15   20M

SECTION
剖面

205

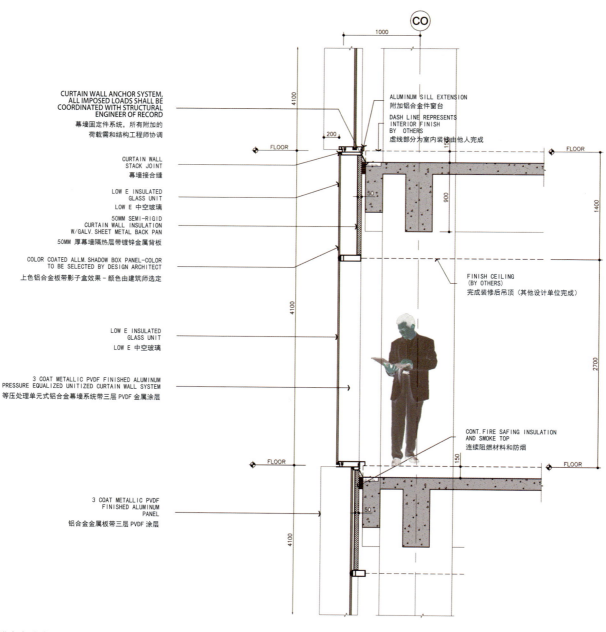

CURTAIN WALL ANCHOR SYSTEM, ALL IMPOSED LOADS SHALL BE COORDINATED WITH STRUCTURAL ENGINEER OF RECORD
幕墙固定件系统，所有附加的荷载需和结构工程师协调

CURTAIN WALL STACK JOINT
幕墙接合缝

LOW E INSULATED GLASS UNIT
LOW E 中空玻璃

50MM SEMI-RIGID CURTAIN WALL INSULATION W/GALV. SHEET METAL BACK PAN
50MM 厚幕墙隔热层带镀锌金属背板

COLOR COATED ALLM. SHADOW BOX PANEL-COLOR TO BE SELECTED BY DESIGN ARCHITECT
上色铝合金板带影子盒效果－颜色由建筑师选定

LOW E INSULATED GLASS UNIT
LOW E 中空玻璃

3 COAT METALLIC PVDF FINISHED ALUMINUM PRESSURE EQUALIZED UNITIZED CURTAIN WALL SYSTEM
等压处理单元式铝合金幕墙系统带三层 PVDF 金属涂层

3 COAT METALLIC PVDF FINISHED ALUMINUM PANEL
铝合金金属板带三层 PVDF 涂层

ALUMINUM SILL EXTENSION
附加铝合金件窗台

DASH LINE REPRESENTS INTERIOR FINISH BY OTHERS
虚线部分为室内装修由他人完成

FINISH CEILING (BY OTHERS)
完成装修后吊顶（其他设计单位完成）

CONT. FIRE SAFING INSULATION AND SMOKE TOP
连续阻燃材料和防烟

低辐射镀膜中空玻璃（又称 Low-E 玻璃）不仅具有极为优良的节能性，还具有多种颜色的装饰性效果，节能性体现在其对阳光热辐射的遮蔽性——即隔热性，对暖气外泄的阻挡性——即保温性两个方面。
中空玻璃是由两片或多片玻璃，用内部充满高效分子筛吸附剂的铝框间隔出一定宽度的空间，边部再用高强度密封胶密封粘合而成的玻璃组件。
中空玻璃内的密封空气，在铝框内灌充的高分子筛吸附剂作用下，成为导热系数很低的干燥空气，从而构成一道隔热、隔音屏障。

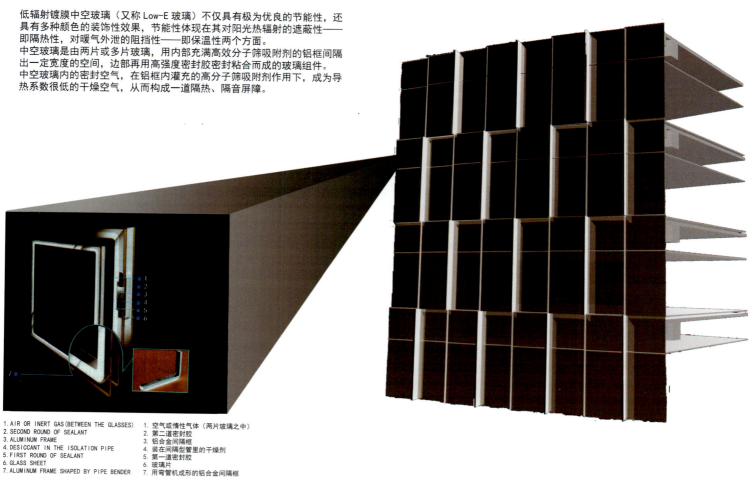

1. AIR OR INERT GAS (BETWEEN THE GLASSES)　1. 空气或惰性气体（两片玻璃之中）
2. SECOND ROUND OF SEALANT　2. 第二道密封胶
3. ALUMINUM FRAME　3. 铝合金间隔框
4. DESICCANT IN THE ISOLATION PIPE　4. 装在间隔型管里的干燥剂
5. FIRST ROUND OF SEALANT　5. 第一道密封胶
6. GLASS SHEET　6. 玻璃片
7. ALUMINUM FRAME SHAPED BY PIPE BENDER　7. 用弯管机成形的铝合金间隔框

CURTAINWALL DETAIL
幕墙系统说明

▶ Vehicle Entrance to Hotel/
Conference/Restaurant
酒店/会议/餐饮车行入口

▶ Entrance to Hotel/Conference/
Restaurant, Club Main Entrance
酒店/会议/餐饮人行入口, 会所主入口

▶ Office Main Entrance
办公主入口

▶ Bank Entrance
银行入口

GROUND FLOOR PLAN
一层平面图

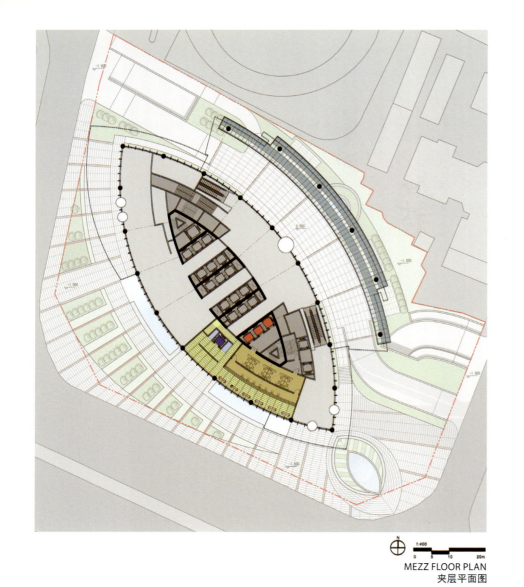

在地下室设计中，经济性是设计师设计的初衷，设计师创造性地使用了斜板式停车，采用了无楼层概念，没有专门的机动车道，设计师从 -1F 到 -6F 全采用小于 5% 的坡道式停车节省了机械式停车，为业主节约了大量的设备经费，也节约了每层约 50 米的车道，现车位为 39.1m²/ 个，有较好的经济性。同时达到机动车停车位有 472 个，完全满足了整个塔楼的停车需求。

Economics is the original intention of the basement design. Architects creatively adopt the non-floor concept by cantboard parking. There is no exclusive motorway. From -1F to -6F, all use cantboard parking which is less than 5%, which saves space for the whole mechanical parking and reduces costs for the owner. Each floor saves 50 meters' driveway , 39.1m²/stall with better economy. There are 472 stalls in total, which can meet the parking demands for the whole building.

MEZZ FLOOR PLAN
夹层平面图

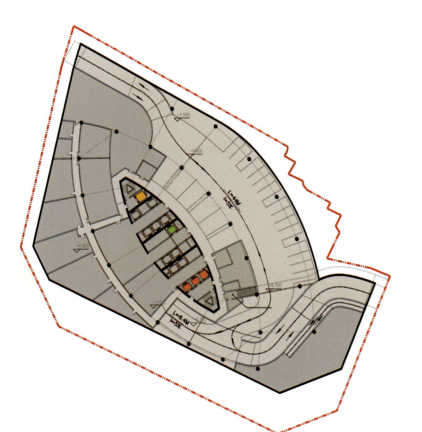

BASEMENT 1 FLOOR
地下一层平面图

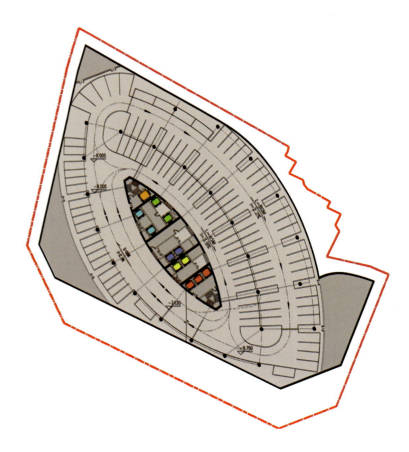

BASEMENT 2 FLOOR
地下二层平面图

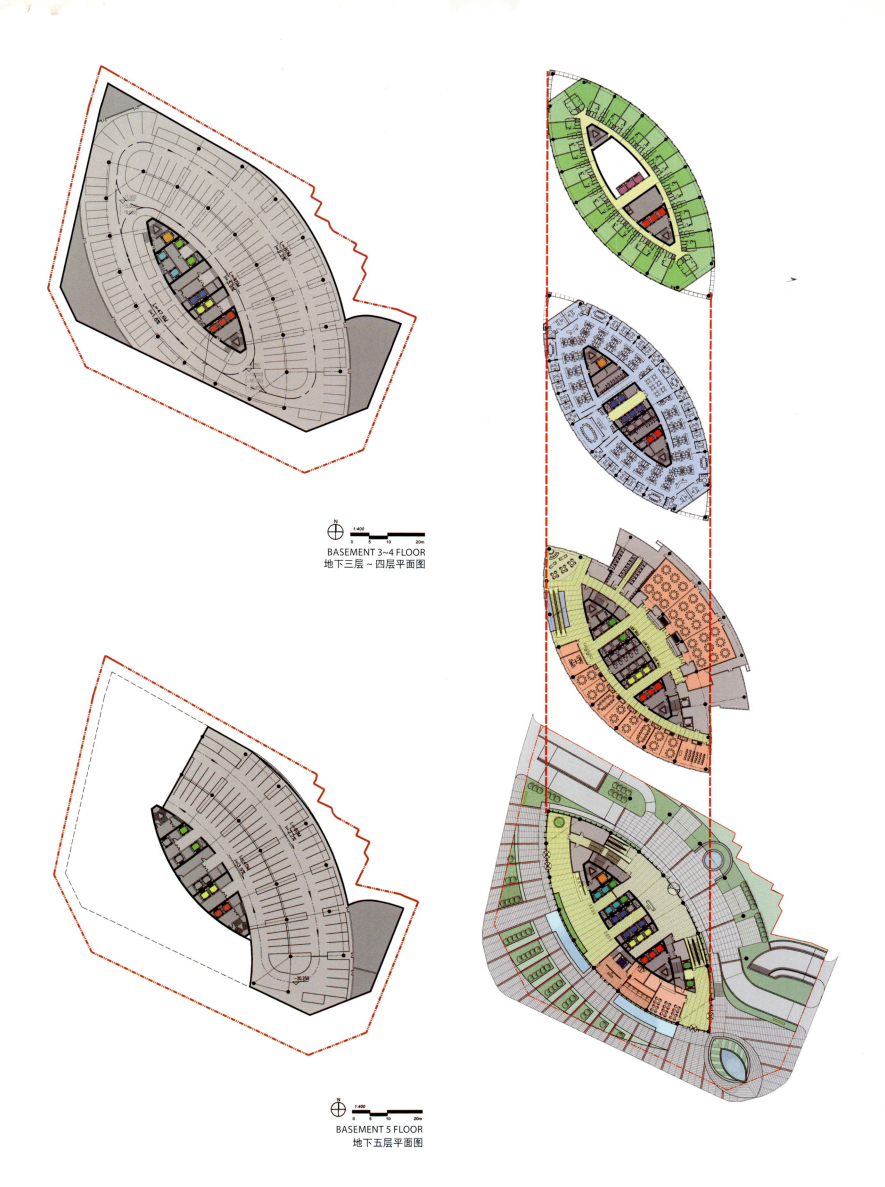

BASEMENT 3~4 FLOOR
地下三层 ~ 四层平面图

BASEMENT 5 FLOOR
地下五层平面图

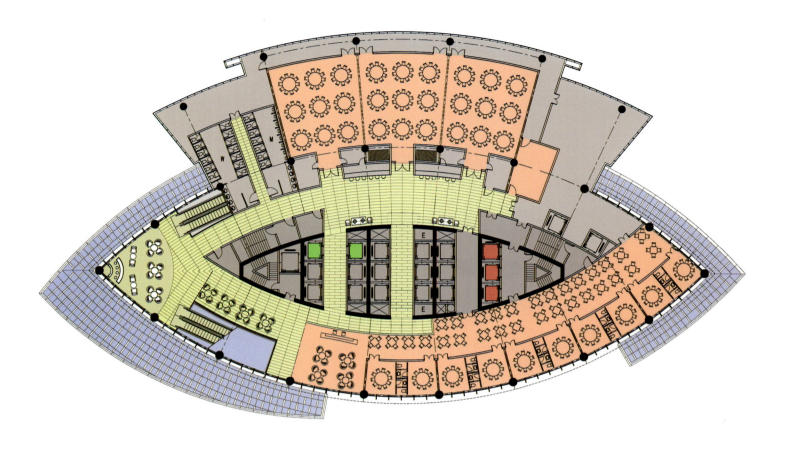

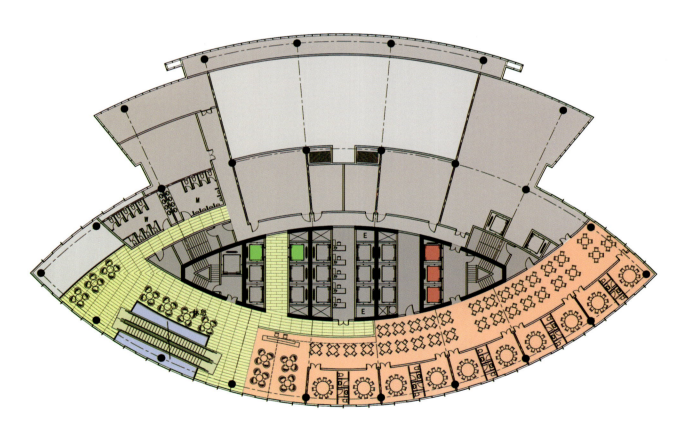

2ND~3RD FLOOR-RESTAURANT
2~3 层 – 餐厅

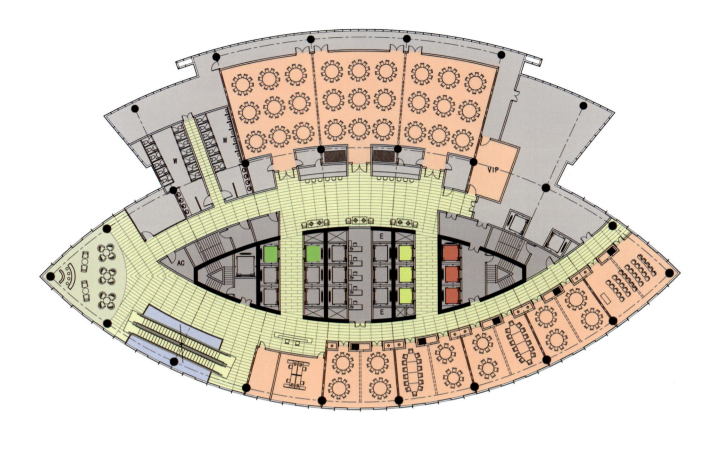

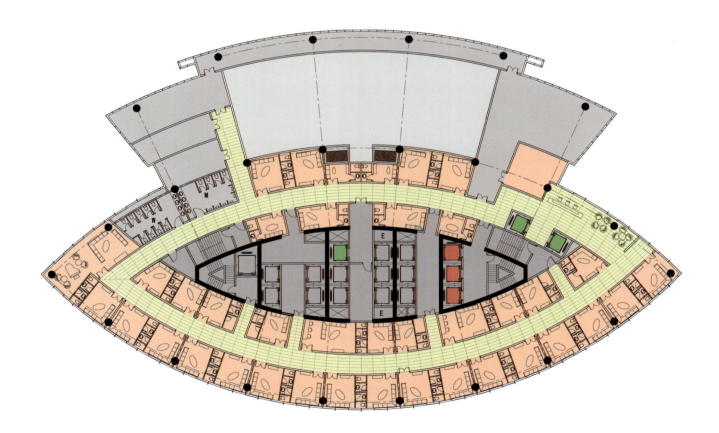

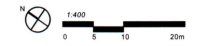

4~5TH FLOOR-MEETING/CLUB

4~5 层 – 会议中心 / 会所

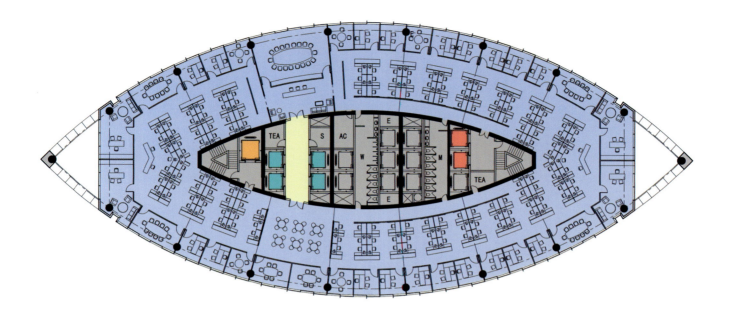

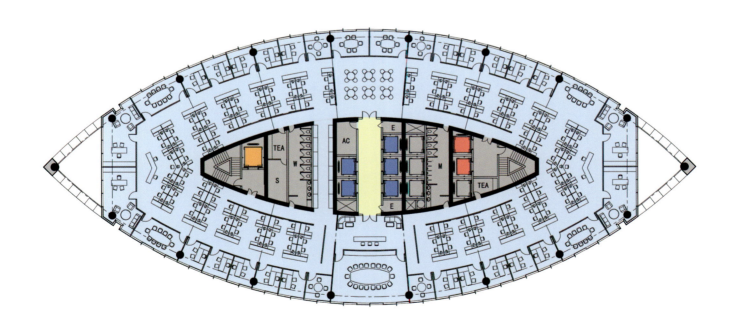

TYPICAL OFFICE PLAN -1
标准办公层平面图 -1

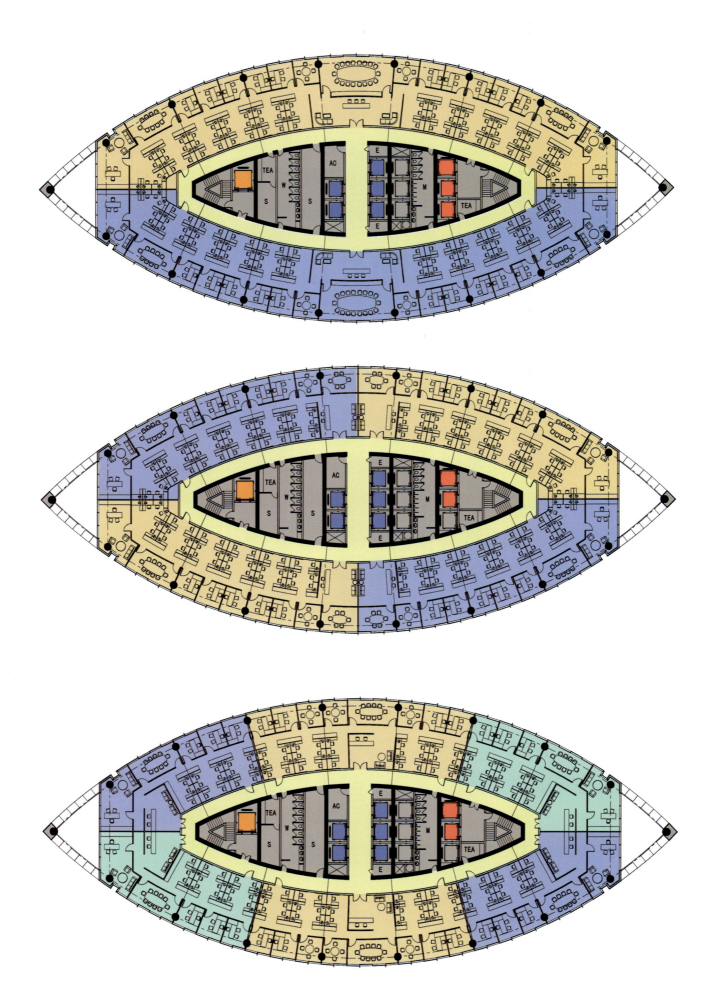

TYPICAL OFFICE PLAN -2
标准办公层平面图 -2

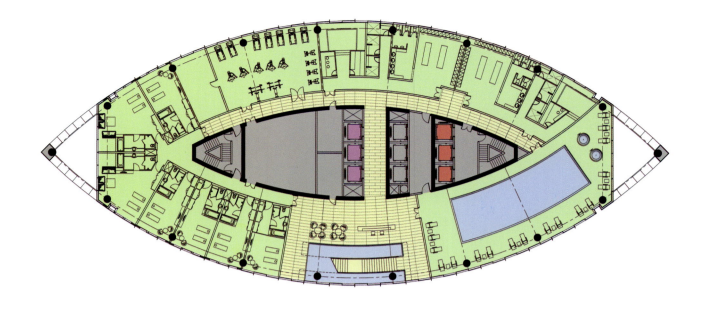

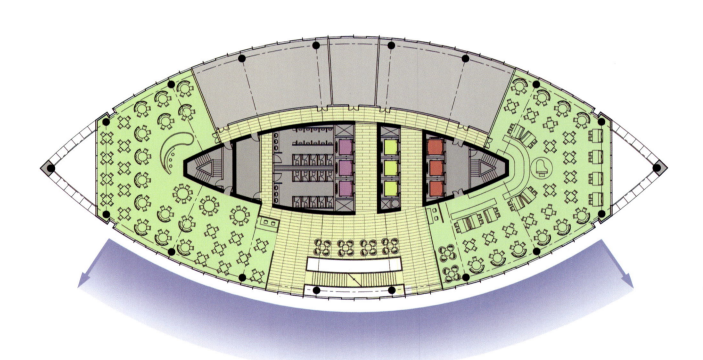

RESTAURANT/SPA/FIYNESS
餐厅 /SPA/ 健身房

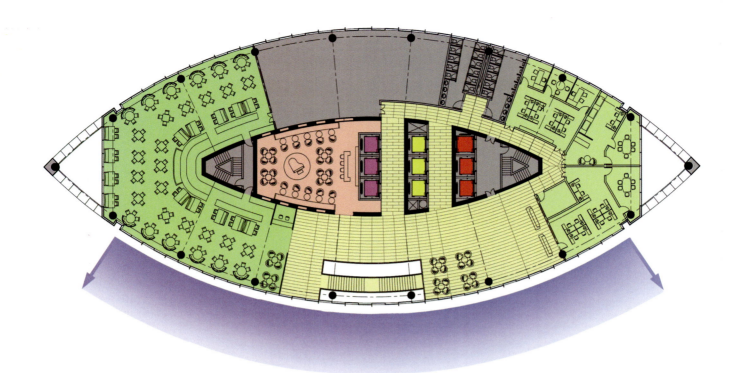

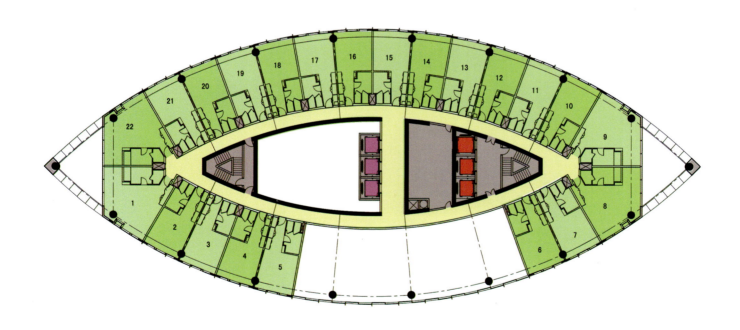

HOTEL SKYLOBBY/GUESTROOM FLOORS
酒店大厅、客房层平面图

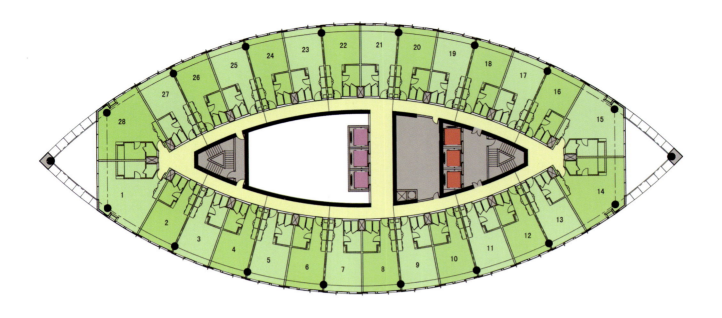

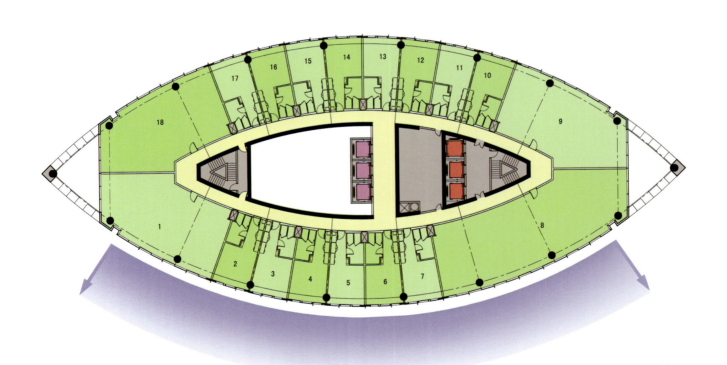

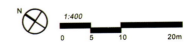

36~47TH FLOOR-HOTEL GUEST ROOM
36~47 层 - 酒店客房平面图

NEW KAGRAN CENTER—DISTRICT LANDMARK

新卡根中心

LOCATION:
VIENNA, AUSTRIA

ARCHITECT: ALLESWIRDGUT
CLIENT: WIRTSCHAFTSAGENTUR WIEN
AREA : 121.950 m²

Around the urban traffic hub, the urban center of the Vienna's 22nd district emerges. The New Kagran Center is supposed to vitalize the neighborhood with fresh architectural esprit and to provide a local landmark. The complex consists of interconnected volumes situated closely together:seven buildings grouped around a common center – the Forum Kagran – like in a village. The compact building elements do not appear anonymous, but with their clear silhouettes have great recognition value. All houses are accessible from inside (the Forum and passageways) and from outside (the square and adjacent streets).

In the central Forum pedestrian traffic flows come together like in a "living room" and as a lively continuation of the outdoor spaces. Tree islands vivify the central square, and generous open spaces provide room for events to be held here. A varied walkway system passes through the structure both in north-south and east-west direction. A landmark building that facilitates and accommodates urban life in all its variety.

SITE PLAN

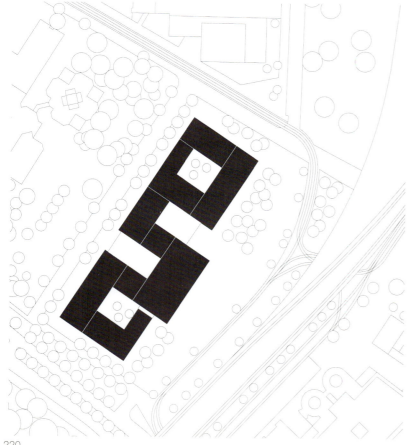

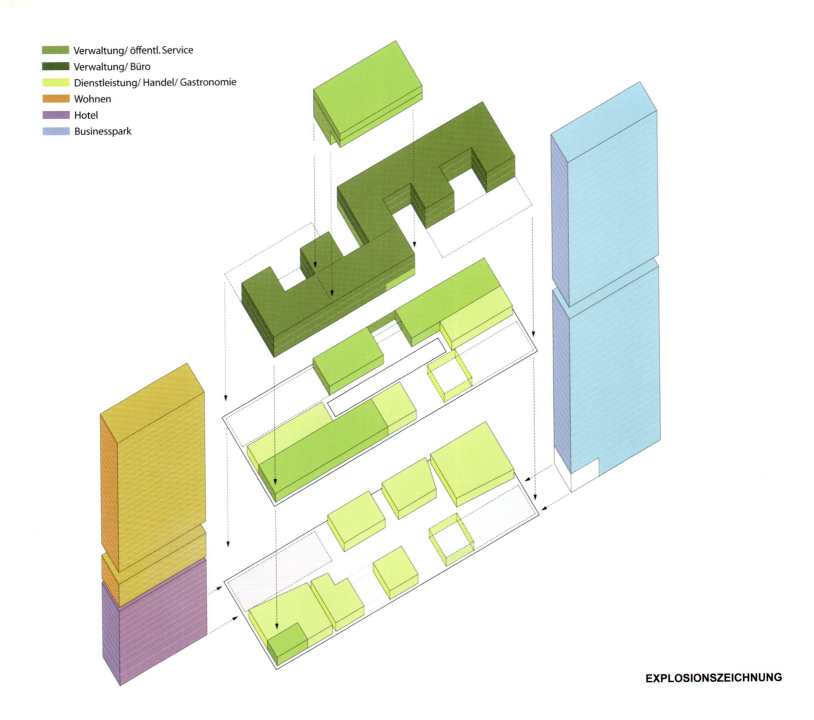

- Verwaltung/ öffentl. Service
- Verwaltung/ Büro
- Dienstleistung/ Handel/ Gastronomie
- Wohnen
- Hotel
- Businesspark

EXPLOSIONSZEICHNUNG

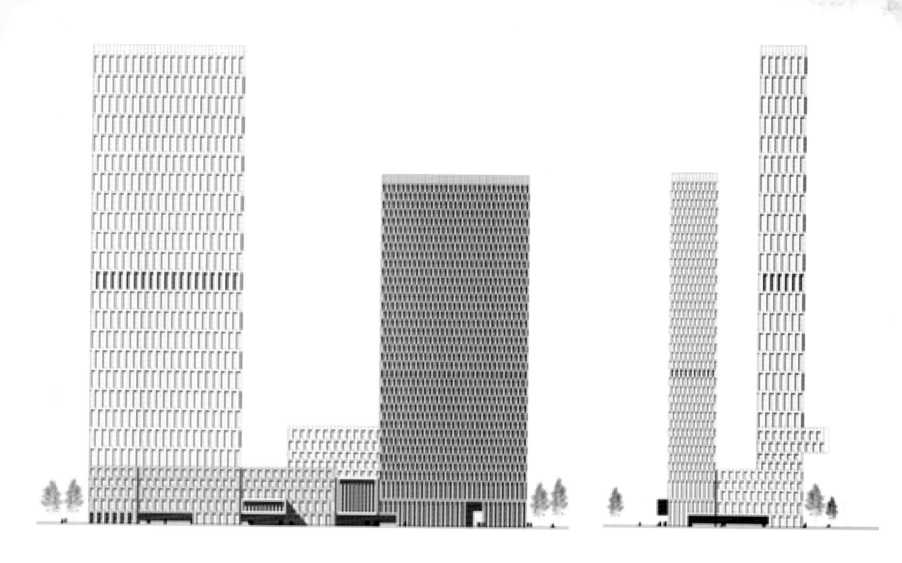

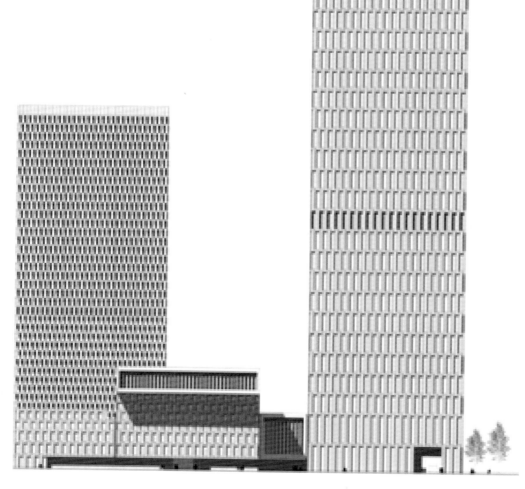

大厦就在城市交通枢纽附近，也成为了维也纳第22区的新的繁华地带。这座新卡根中心将赋予周边地段新鲜的建筑气息，并成为当地的一座地标。这座大厦包括多个错落有致、互相毗邻的空间：7个建筑围绕着"卡根论坛"这个中心建筑，就像一个村庄一样。如此密集的建筑元素并没有让其失去特色，相反，清晰的建筑轮廓具有非常明显的识别度。所有的楼宇内部都是相通的（通过中心的卡根论坛和走廊互通），当然外面也是可以到达的（通过广场和相邻的街道）。

在中央的论坛里，人们聚集在一起，就像来到了一间"客厅"，这里还是室外空间的一种延续，充满活力。中心广场的树岛生机无穷，开阔的空间也成为了举办活动的场所。在南北和东西两个方向都有多变的步行通道。这座地标式的建筑正在用它多变的风格为城市生活提供着便利与服务。

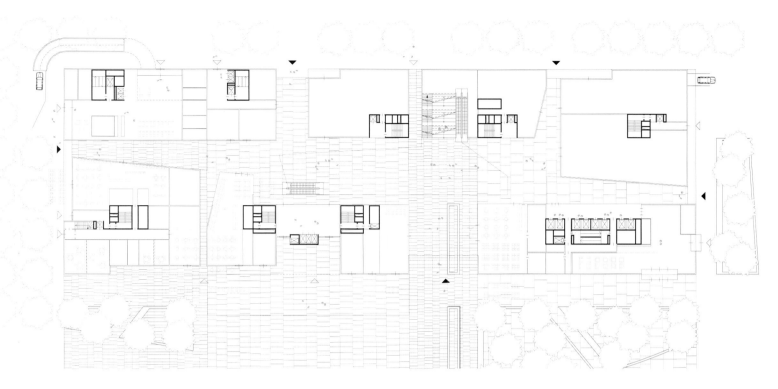

FLOOR PLAN

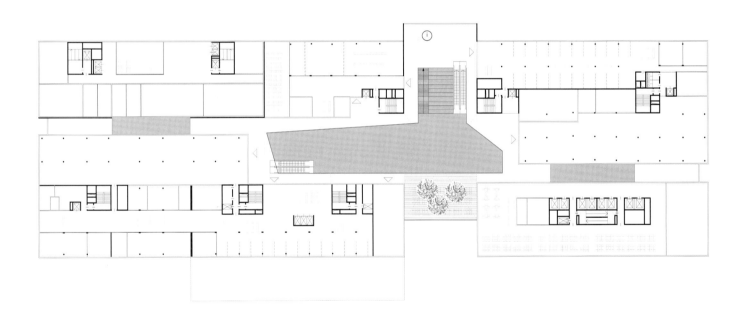

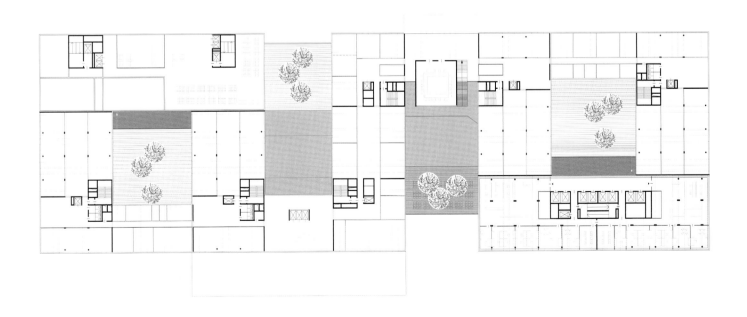

353 N CLARK

克拉克北街 353 号大廈

LOCATION:
CHICAGO, USA

ARCHITECT: LOHAN ANDERSON LLC
ARCHITECT OF RECORD: A. EPSTEIN & SONS INTERNATIONAL
MEP/FP/TELECOM ENGINEER: ENVIRONMENTAL SYSTEM DESIGN

LANDSCAPE ARCHITECT: DANIEL WEINBACH & PARTNERS
CLIENT: MESIROW STEIN REAL ESTATE
DEVELOPER: SOUTH PARCEL DEVELOPMENT, LLC

CONTRACTOR: BOVIS LEND LEASE
AREA: 1,438,265 m²
PHOTOGRAPHER: BILL ZBAREN

The high-rise office tower was designed with input from both tenants to develop a custom-designed business solution to their respective facilities' requirements. Beyond the architectural expression, design criteria demanded by the tenants included: efficient mechanical systems that maintain a comfortable environment for the tenants while reducing energy costs; proven efficiency of floor plate size and open-to-closed office space ratios; high level of security for tenant safety and business protection without disrupting client interaction; infrastructure for technology maximization and adaptability for future advances.

The exterior of the building is sheathed in Low-E glass, some fritted glass and aluminum spandrel panels. These panels are concave horizontal bands surrounding each floor giving the building a distinctive horizontal layering appearance. Office floors are not complete rectangular shapes but are articulated with several setbacks on the perimeter to accentuate the verticality of the 45 floors. At each setback, a free floating glass panel shoots out into space covering the exposed columns.

The building responds to its unique location close to the Chicago River where the adjoining streets are ramping up to bridges. The slope is unique in Chicago, which is otherwise completely flat, and offered an opportunity to create a private and elevated approach driveway between two streets.

The tower is setback from Clark Street to the West, providing a sizeable plaza that is landscaped and a welcome relief to the daily life of the surrounding neighborhood.

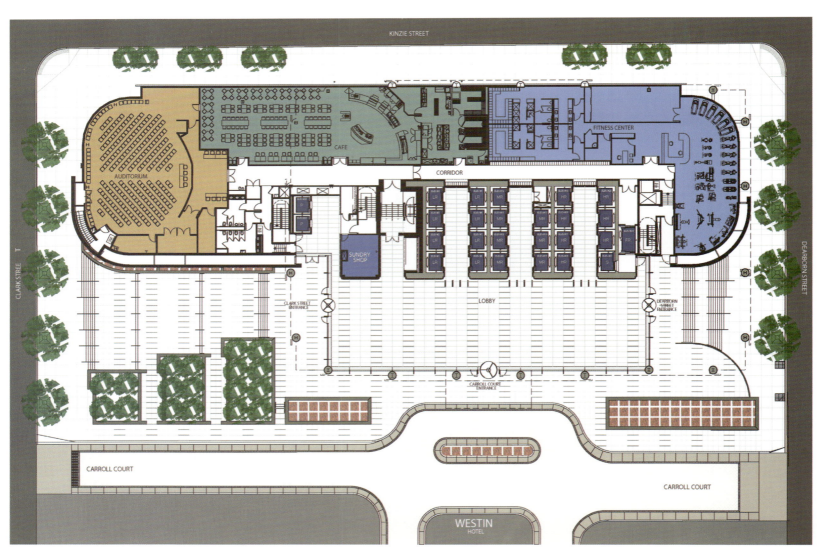

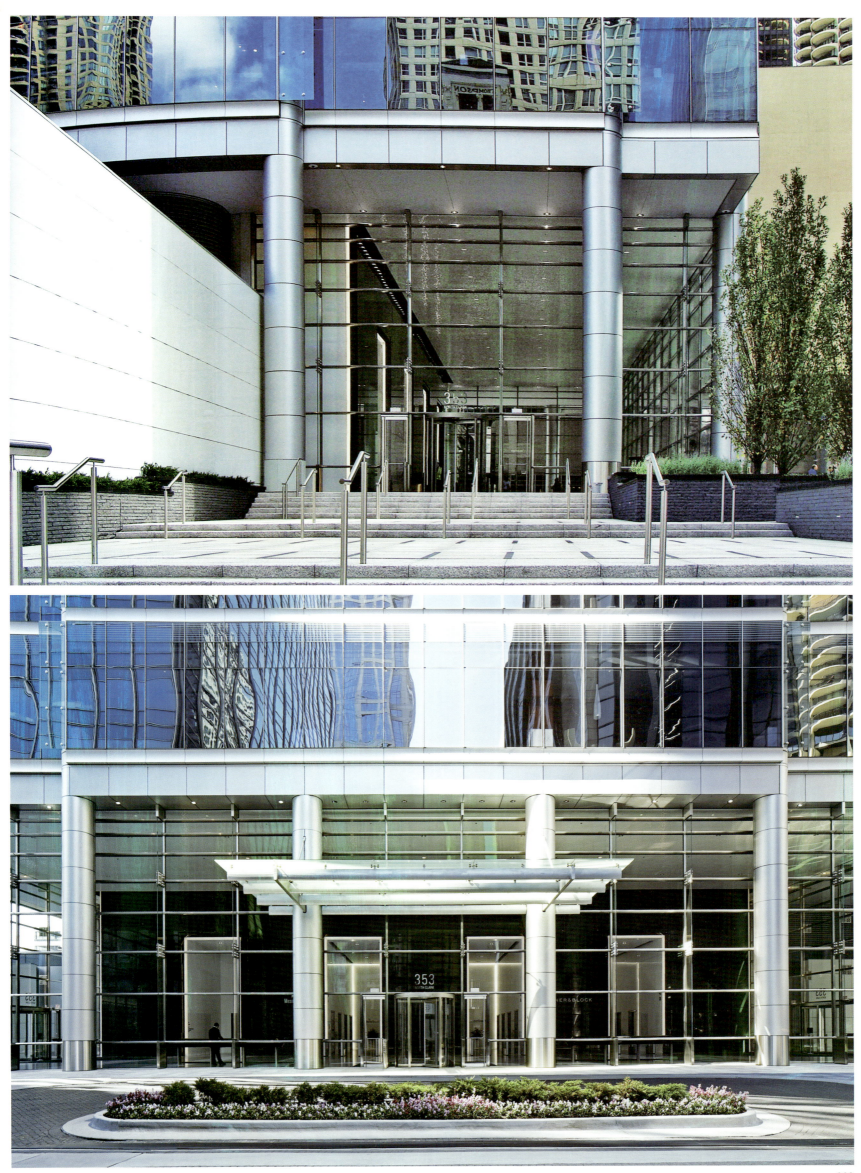

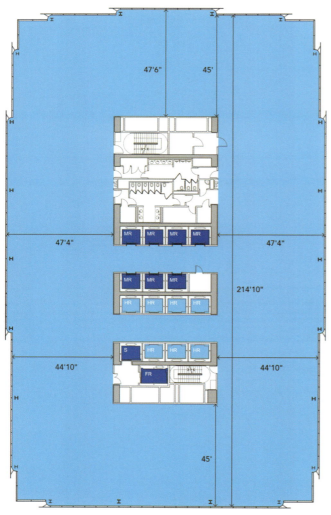

高层办公塔楼的设计，需要满足未来租户因为各自不同的商业活动而对设施提出的要求。除了建筑表达外，租户要求的设计标准还包括：能够维持租户环境舒适而且能源消耗较低的有效的机械系统，高效的楼面板尺寸及合理的开放－封闭办公空间比率，确保租户安全和商业活动得到保护，同时不影响客户间的互动；能够最大程度地适应未来技术升级的基础设施。

大厦外部安装的是低辐射玻璃，间隔一些烧结玻璃和铝框的拱肩镶板。镶板凹进每层的楼面中，使得大厦呈现出独特的水平分层的外观。每个楼层的形状都不完全是矩形的，而是在周边铰接了几个退缩角，以突出这座45层高楼的垂直度。在每个退缩角处，安装一个独立的浮动玻璃嵌板，并向外伸出盖住裸露的柱子。

该建筑地理位置独特，靠近芝加哥河，同时毗邻的街道地势上升与跨河大桥相连，形成了一个斜坡。这个斜坡在芝加哥是独一无二的，借此可以在两条街道之间建设一条专用高架车道。

大厦西侧距克拉克街有段距离，因此在街道与建筑中间形成了一个颇具规模的广场，为周边的群众提供了一个日常休闲购物的场所。

TYPICAL MID-RISE FLOOR PLAN

MENARA BINJAI

宾甲大厦

LOCATION:
KUALA LUMPUR, MALAYSIA

ARCHITECT: LILLIAN TAY, JACKY LEE, ISKANDAR RAZAK, LAU KOK SAN
AREA: 85,000 m²

PHOTOGRAPHER: PAUL GADD

Originally the site of a private family home, Menara Binjai is an office tower development conceived to maintain the viability of keeping the plot of land within the family. The green impetus manifests in a series of triple-volume sky gardens incorporated into the front facade of the building that ascend to culminate in the greenery of the roof garden.

A permeable glass screen creates "a breathable skin" in the curtain wall, allowing prevailing north-westerly breezes to infiltrate the sky gardens whilst buffering the terrace space from winds and heavy monsoon rain. This "urban verandah" allows building users to experience the tropical setting, a contrast to the typical contained office environment.

A canopy of custom-designed timber louvres serves as a brise-soleil creating a dramatic play of light and shadow on the roof terrace. The louvres are proportioned in scale with the glass wall to achieve a balance between shelter and openness to the tropical climate, creating a comfortable micro-climate for this semi-outdoor space.

The ascending sky gardens becomes a strong compositional element of the main facade and a counterpoint to the sleek low-emission double-glazed curtain wall, which enhances the thermal performance of the building envelope. Deep vertical fins on the east and west facade accentuate the verticality of the building and multi-tasks as passive sun-shading devices that modulates daylight, without cutting out views. This helps to maximize the energy performance of the building without compromising on views.

In a symbolic gesture of continuity and a deliberate invocation of the site's history, the mahogany trees which stood in the garden of the old family home, were saved during demolition and the timber was harvested to be re-used in the building as a feature wall in the reception, the lobby seating and for lift finishes – a quiet celebration of the memory of the site.

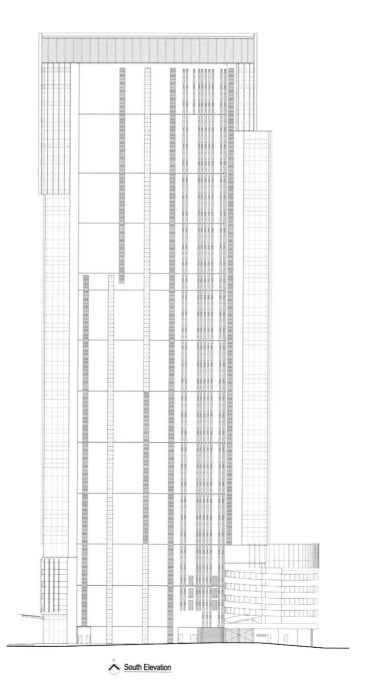

△ N South Elevation

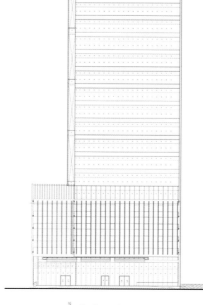

△ N Pandangan Barat

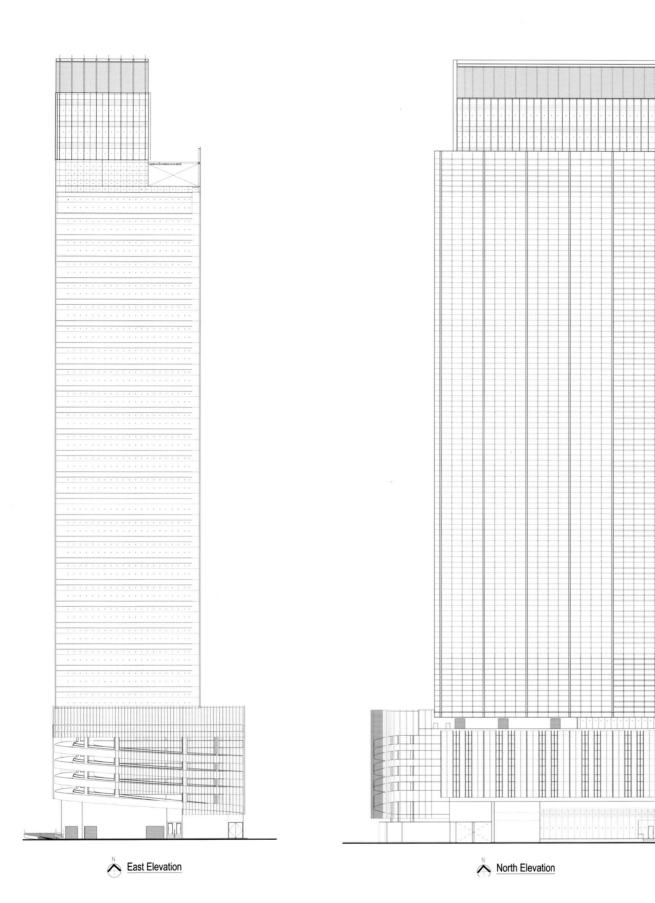

N East Elevation

N North Elevation

最初只是一个私人住宅，后来，宾甲大厦成为了一个办公大楼项目，以期望能保持这块住宅地块的生命力。大厦的绿色概念体现在大厦正立面顶层的一系列三维空中花园。这些绿色的花园向上攀升，与顶层花园融为一体。

大厦采用可渗透玻璃幕墙，就像一层可以呼吸的皮肤一样，让盛行的西北方向的微风吹过空中花园，使门廊免遭大风雨的侵袭。这种"都市门廊"让大厦里的人们体会到了热带气候，与传统的封闭式办公环境形成鲜明对比。

顶棚上有专门定制的木质百叶窗，充当了遮阳器的角色，在屋顶平台上产生了奇妙的光影效果。百叶窗的尺寸与玻璃幕墙比例相宜，使屋内的光线与阴影达到平衡，为这座半露天的空间创造了一个舒适的微环境。

这些向上延伸的空中花园成为外立面的一个鲜明的结构元素，完善了光亮的低辐射双层幕墙，成为大厦恒温系统的重要设施。东、西外立面采用深长的垂直鳍设计，突出了大厦的纵深感，也承担了遮阳的功能，还不影响视野。这样的设计最大限度地强化了整个建筑的节能效果，同时还不影响视野。

最具象征和延续意义的设计是在拆迁的过程中得以完整保留的那些长在老宅花园里的桃花心树，其木材被用到了大厦的接待处、大厅座椅和电梯装饰中，成为了大厦的一个特色，这也是在默默地在向历史致敬。

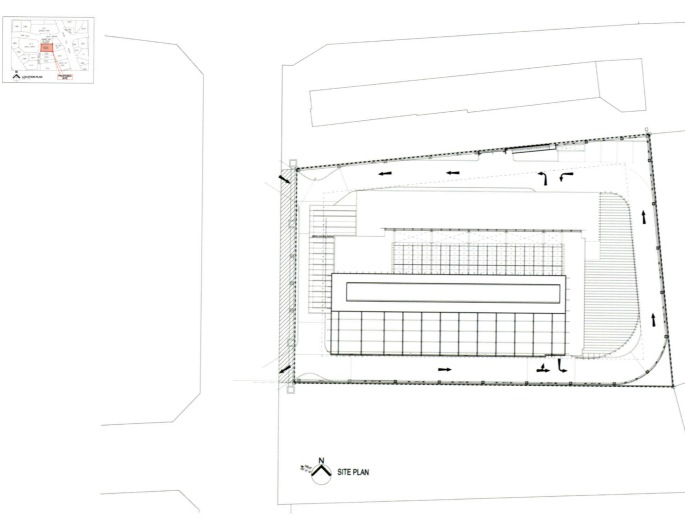

SITE PLAN

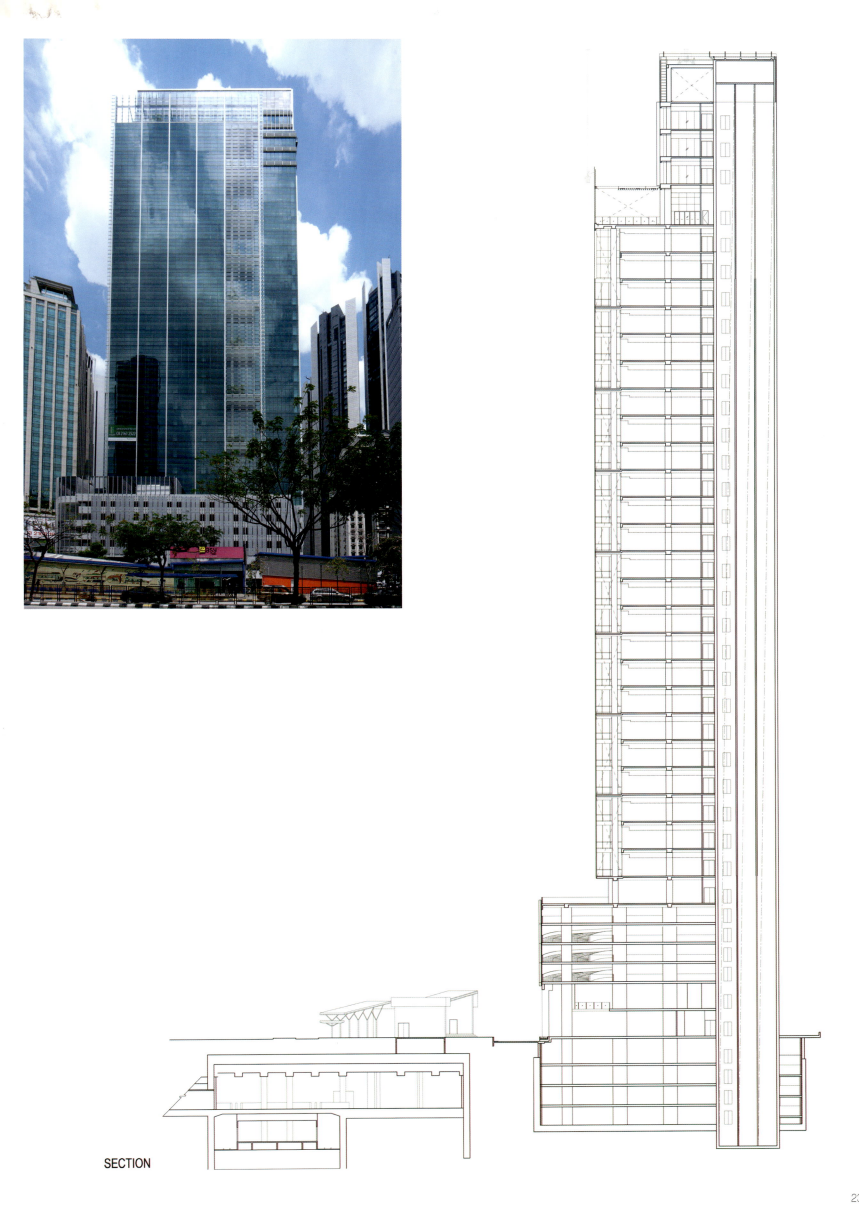

SECTION

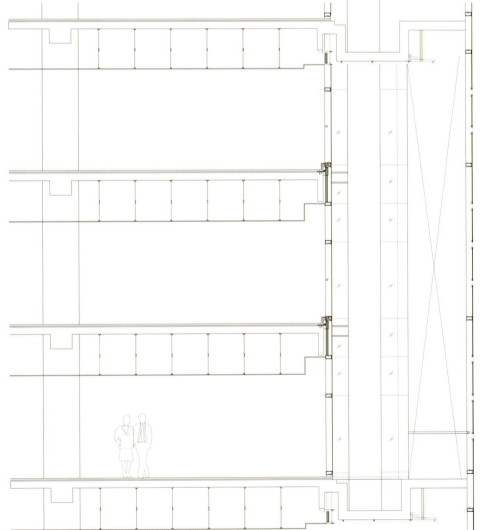

SECTION OF SKY GARDEN

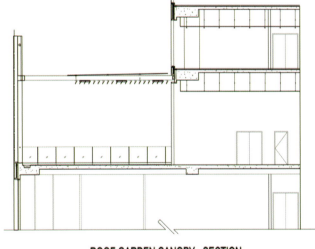

ROOF GARDEN CANOPY—SECTION

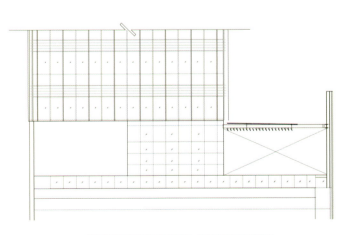

ROOF GARDEN CANOPY—EAST ELEVATION

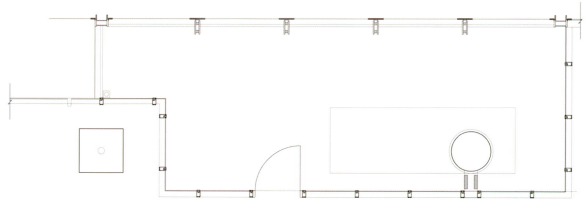

PLAN OF SKY GARDEN

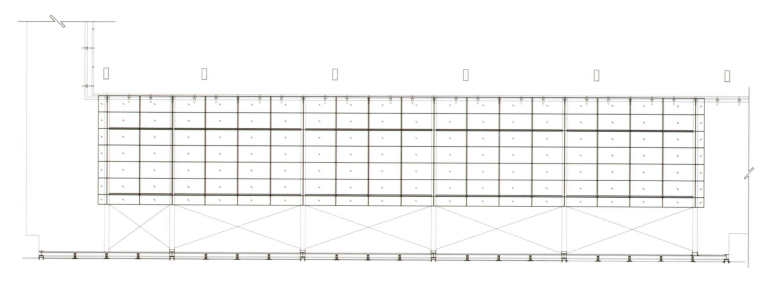

ROOF GARDEN CANOPY PLAN

HARMONY TOWER, SEOUL, KOREA

和谐大厦

LOCATION:
SEOUL, KOREA

ARCHITECT: STUDIO DANIEL LIBESKIND
SKETCH AND RENDERING DESIGNER: DANIEL LIBESKIND, SDL, CRYSTAL
AREA: 100,000 m²

Harmony Tower, a project that is part of the new Yongsan International Business district (YIBD) development in Seoul, is an iconic, 21st century sustainable office tower that is 46 floors. The design for the tower is inspired by YunDeung, traditional Korean paper lanterns. The concept is to create a tower as a faceted lantern, whose multiple planes reflect the sky and the earth and capture the light on its differing angles, creating a glowing gateway and beacon in the YIBD site.

The sculpted tower is subtly shaped by the urban context. The tower tapers at its base to create a feeling of space and openness for the pedestrian plaza. The form then reaches out in the middle of the tower to maximize the floor plates and Han River views and to create a sense of scale marking the gateway from the western entrance to the site. The tower then tapers back and up to its top to allow the most light and air onto the other towers around, creating a strong ascending peak to the tower. The tower form creates multiple perspectives, like a sculpture in the round, with an ever changing public profile responding specifically to the site.

The tower contains unique vertical winter gardens on the south and west facades, providing users access to natural ventilation and planted park settings at each of the 38 office floors. The gardens not only act as a special amenity to all the tenants, but also a buffer to the direct sunlight hitting the glass building. The gardens function as open, park space within the building, but also help to reduce the heat gain and allow the building to function more sustainably. Harmony Tower is a state of the art workplace, interweaving themes of nature, sustainability, and efficiency in a faceted, sculptural form.

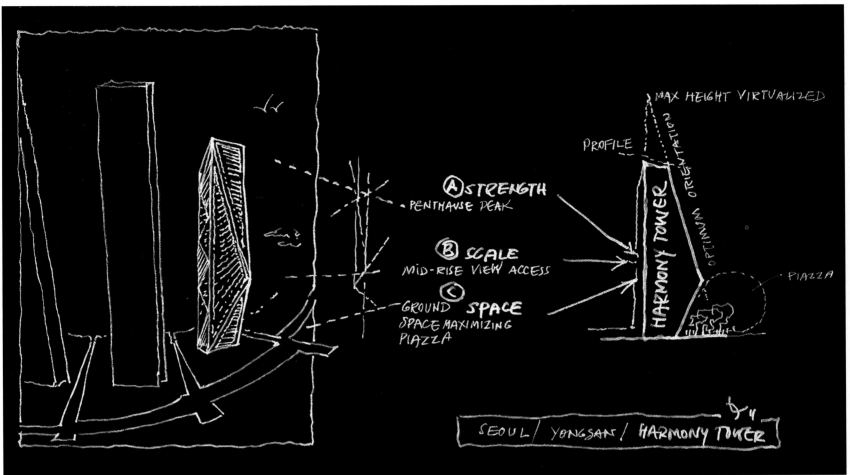

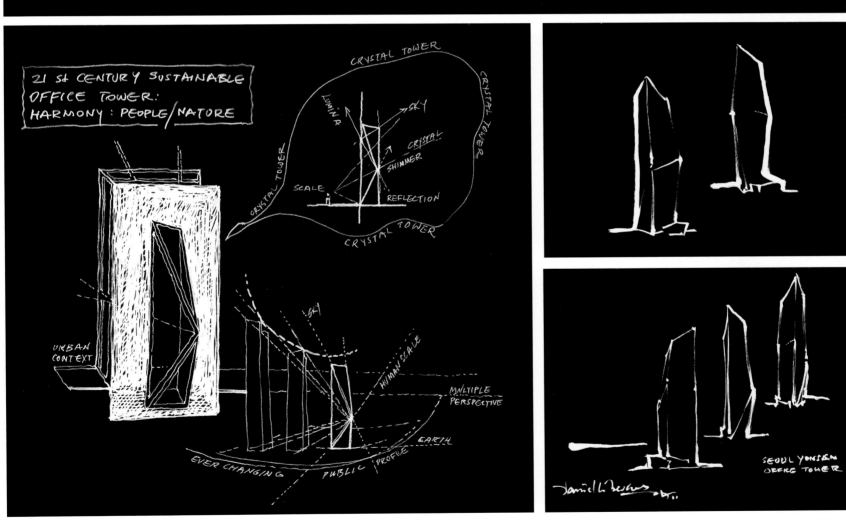

和谐大厦，是在首尔新龙山国际商务区（YIBD）开发的一个项目，是一座标志性建筑，一座21世纪的可持续办公大楼，高46层。大厦设计的灵感来自YunDeung，YunDeung是韩国的一种传统的纸灯笼。其理念是创建一个多面灯笼形状的大厦，大厦的多个平面可以照出天空、大地，并捕获来自不同角度的光线，从而成为YIBD地区闪闪发光的门户和灯塔。

这个具有特色的大厦受城市文脉的巧妙影响。大厦底部呈锥体，以创造步行广场的空间感和宽阔感。大厦外观从中部向外延伸，以获得最大的楼面面积和最佳的汉江景色，并创造一种规模感，从而成为西面商务区的门户。然后，大厦再次逐渐变细，一直到楼顶，以便周边的建筑可以得到更多的阳光和空气，从而营造出一种强烈的"登峰造极"的感觉。大厦的外观使其拥有多个视角，像一个有立体感的雕塑，可根据位置的不同时时变化外观轮廓。

大厦朝南和朝西的立面上有两个独特的垂直冬景花园，均位于写字楼的38层，用户可以前往这个自然通风的种植花园。花园不仅是为所有租户提供的便利设施，也是大厦阻隔直射太阳光的一个缓冲。花园是大厦中的开放公园，有助于降低热增量，确保大厦的可持续性。和谐大厦是一个体现先进设计水平的工作场所，这个有着多面、雕刻般外观的建筑，整体设计交织着自然、可持续性和效率的主题。

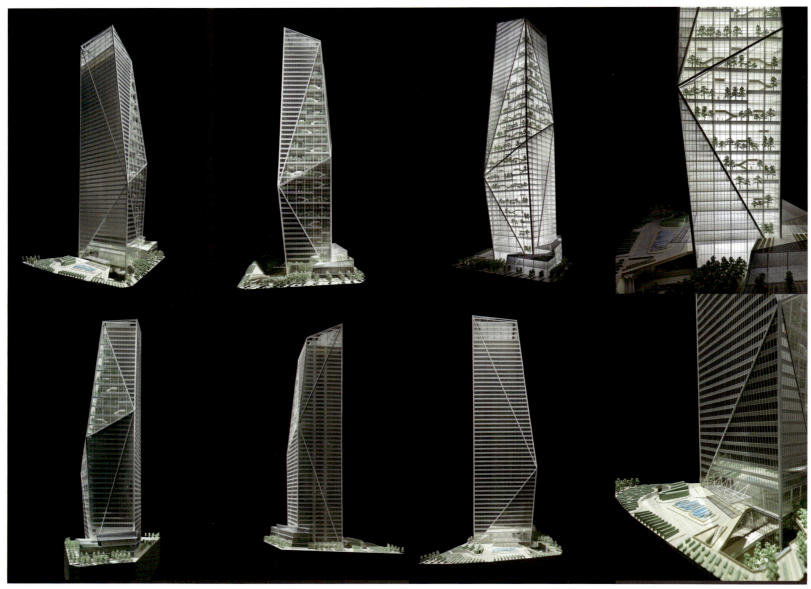

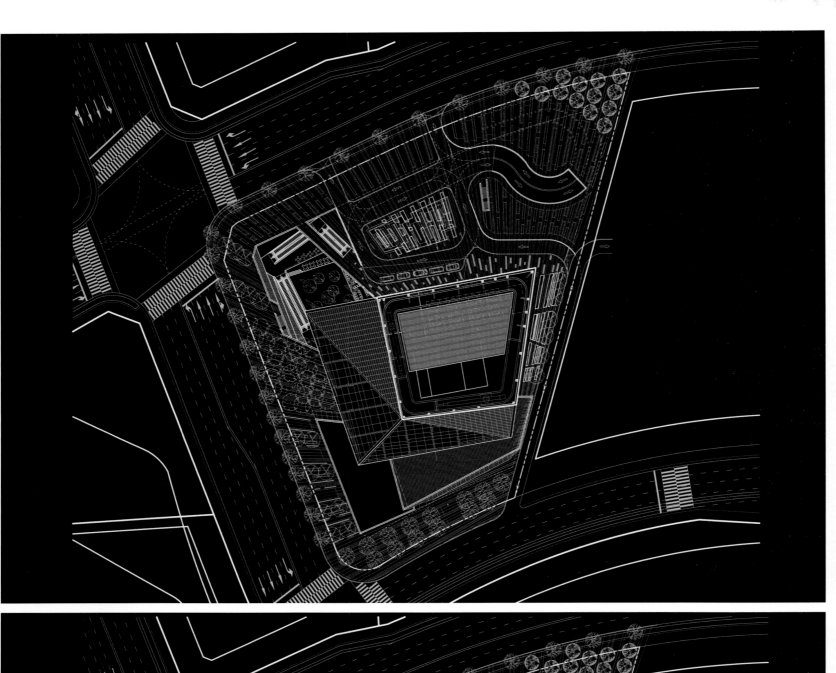

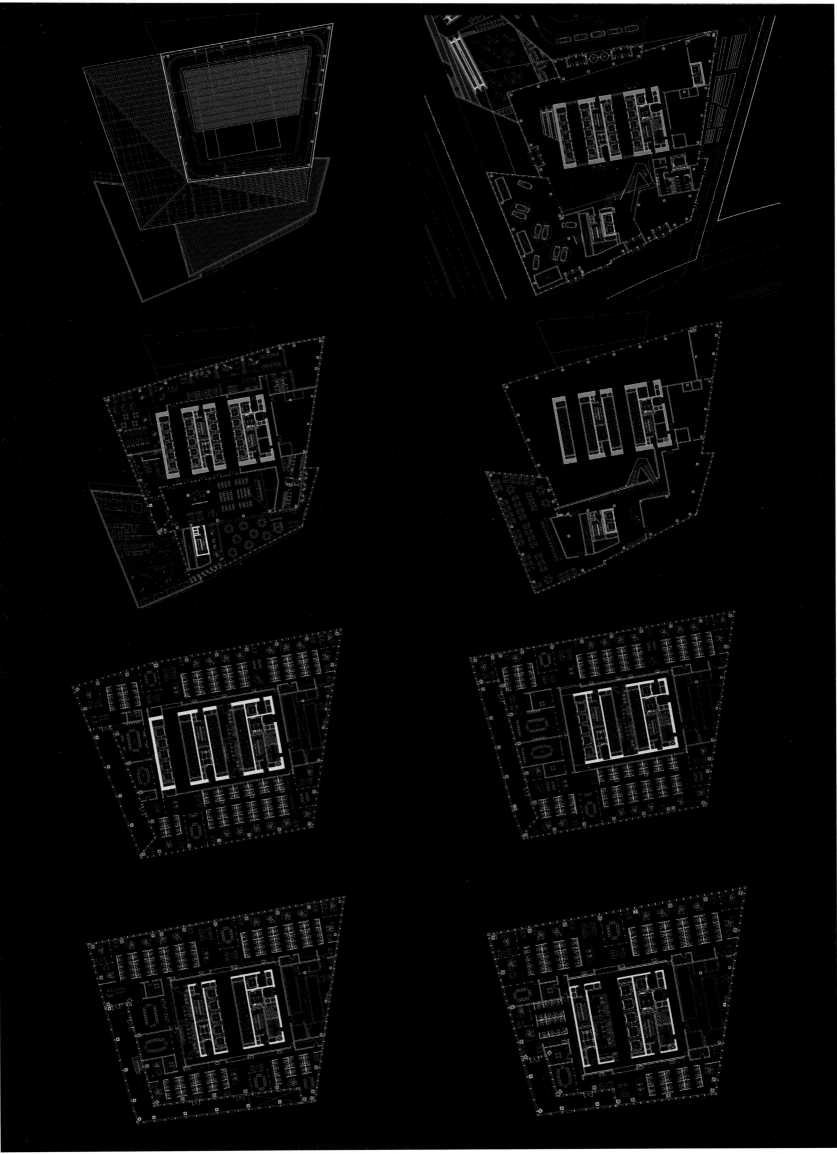

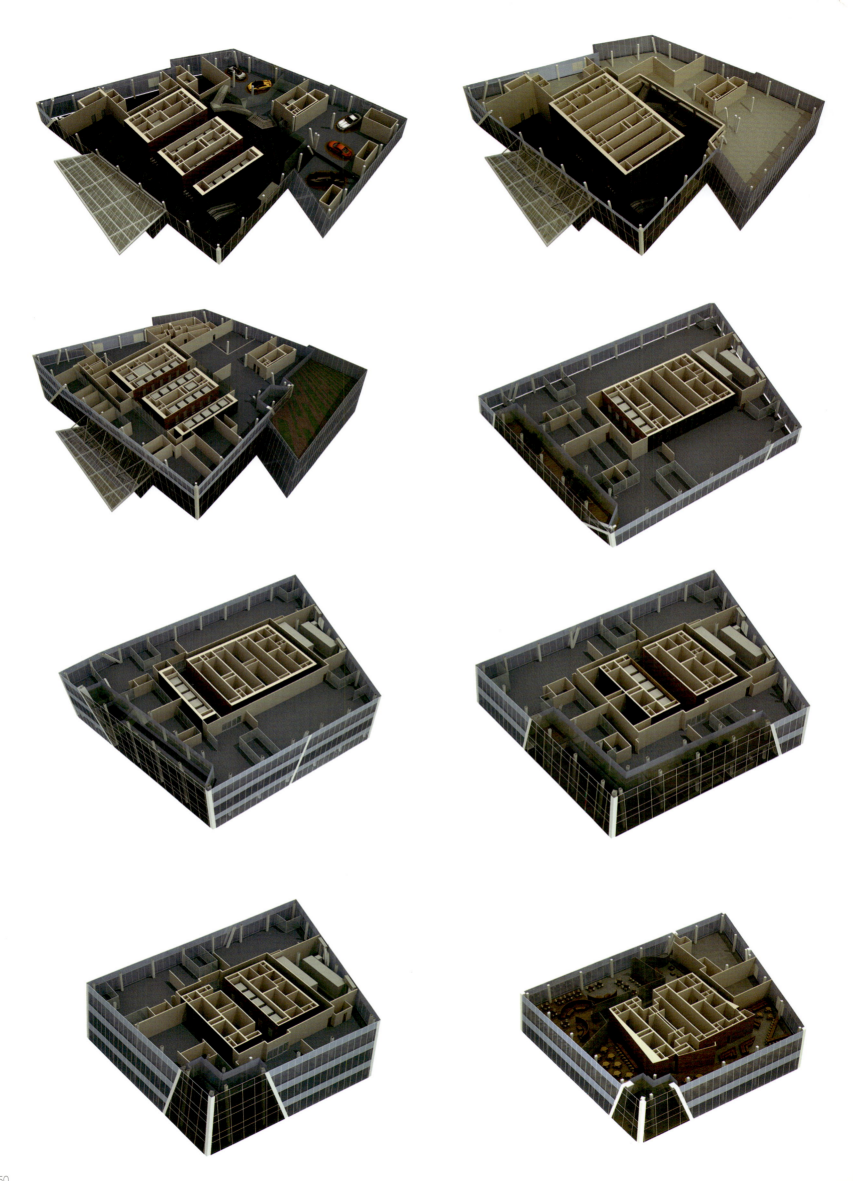

SHIMIZU HEADQUARTERS

清水公司总部大厦

LOCATION:
TOKYO, JAPAN

ARCHITECT: SHIMIZU CORPORATION
AREA: 51,355m²

Shimizu Corporation, one of the largest general contractors in Japan, has completed work on a building that emits the world's least amount of CO_2. The state-of-the-art building emits only 38 kg/m² of CO_2 per year, 62% less on average than ordinary buildings in Tokyo. Shimizu has developed and adopted various technologies to reduce CO_2 emissions. One representative technology is an air conditioning system that makes use of radiant heat. Water hoses run under ceiling boards like capillary vessels. By controlling the temperature of the water circulating in the hoses, the temperature of the celling board surface is controlled. As a result, a surface temperature of about 20 degrees absorbs the heat of people working in the office through a radiant effect.

This system can reduce CO_2 emissions by 30% compared with conventional air conditioning systems. The lighting system also makes use of energy-efficient

technology. Light-emitting diode (LED) lighting is fully adopted and controlled by motion sensors. Energy used for lighting in the daytime is generated by photovoltaic (PV) panels placed in the outer walls. The area of the PV panels is about 2,000m² and generates 84,000 kwh of power per year.

Furthermore, Shimizu has installed window shades that allow sunlight into the offices. The shade angle automatically changes to follow the sun and optimize natural light. These efforts make it possible to reduce CO_2 emissions by 90% compared with standard lighting systems.

By the end of the year 2015, Shimizu will reduce CO_2 emissions down to 70% through fine tuning of air conditioning and lighting facilities as well as adopting further energy saving systems. Finally, Shimizu will offset the remaining emissions by creating emission rights to realize Zero Emission Building (ZEB).

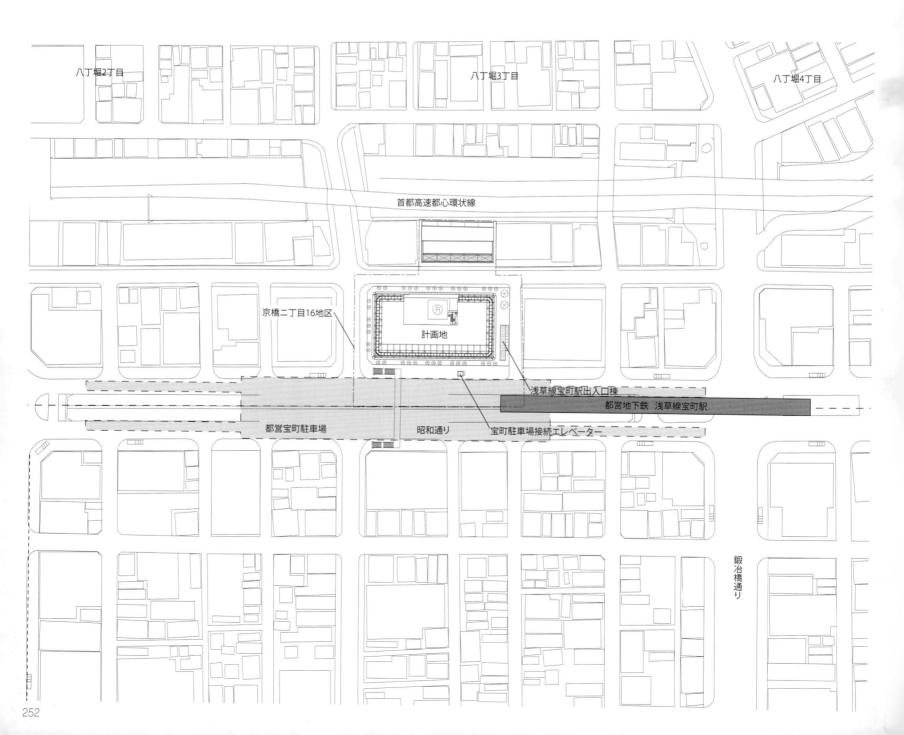

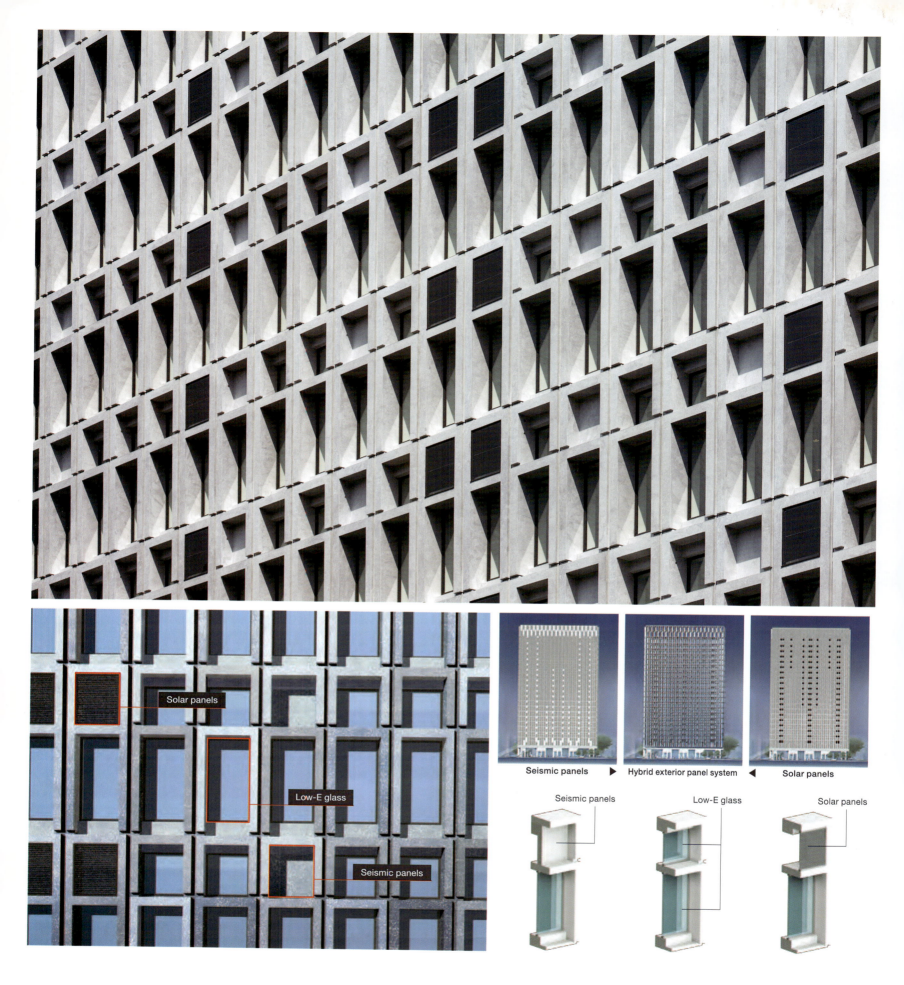

Seismic panels ▶ Hybrid exterior panel system ◀ Solar panels

Seismic panels | Low-E glass | Solar panels

清水公司是日本最大的建筑公司之一，成功建造了世界上最低碳的大楼。这座最先进的大厦每年的二氧化碳排放量为 38 公斤 / 平方米，比日本东京普通的大楼平均少了 62%。清水公司研发和使用了多种技术来降低二氧化碳的排放量。其中一项最具代表性的技术就是使用辐射热能的建筑空调系统。隔板下面的水管像毛细血管一样分布着，通过控制水管的水温，可以控制隔板表面的温度。因此，通过辐射效应，当隔板的温度为 20 度时，它就可以降低办公室中的人的体温了。

与传统空调系统相比，这种系统可以降低 30% 的二氧化碳排放量。大厦的照明系统也使用了节能技术。大厦安装了 LED 照明灯，并通过运动传感器对照明进行控制。白天照明的能源来自于安装在外墙上的太阳能光电板。整座大楼的太阳能光电板面积达 2000 平方米，每年发电 84000 千瓦时。

此外，大厦还安装了遮阳卷帘，可以自动调节进入办公室的阳光，通过角度调控来优化自然光照。这些技术使得大厦的二氧化碳排放量比标准的照明系统的二氧化碳排放量降低了 90%。

到 2015 年年底，大厦将通过调节空调和照明设施，以及更加先进的节能系统，将二氧化碳排放量降低到现在的 70%。最终，大厦将彻底停用和替换现有的排放技术，实现二氧化碳零排放。

Seismic-isolation system

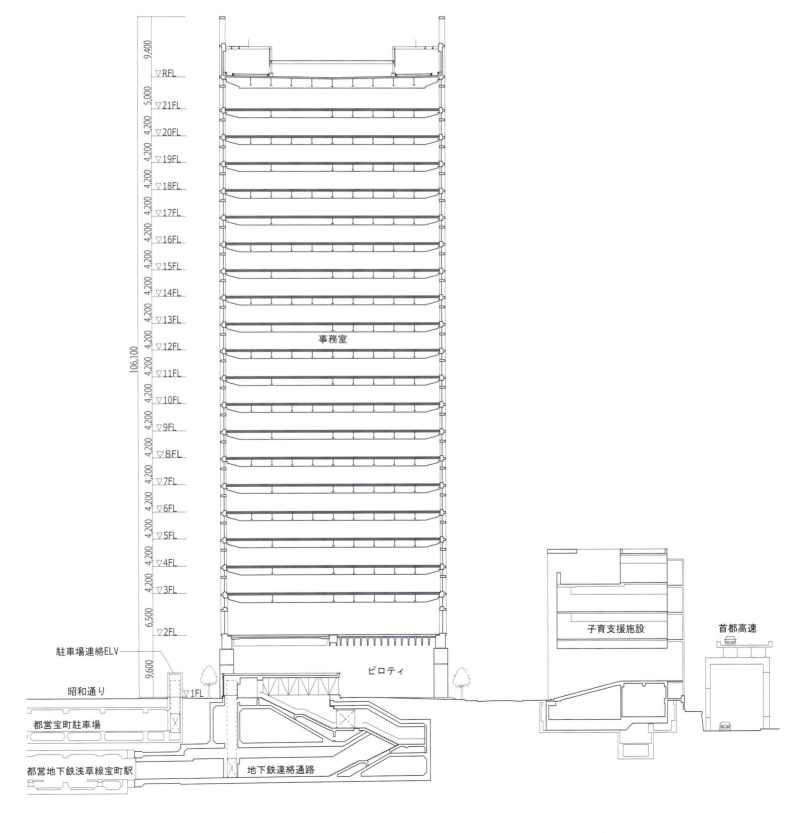

9,400

▽RFL
5,000
▽21FL
4,200
▽20FL
4,200
▽19FL
4,200
▽18FL
4,200
▽17FL
4,200
▽16FL
4,200
▽15FL
4,200
▽14FL
4,200
▽13FL
4,200
▽12FL
4,200
▽11FL
4,200
▽10FL
4,200
▽9FL
4,200
▽8FL
4,200
▽7FL
4,200
▽6FL
4,200
▽5FL
4,200
▽4FL
4,200
▽3FL
6,500
▽2FL

106,100

事務室

子育支援施設

首都高速

駐車場連絡ELV

9,600

昭和通り

▽1FL

ピロティ

都営宝町駐車場

都営地下鉄浅草線宝町駅

地下鉄連絡通路

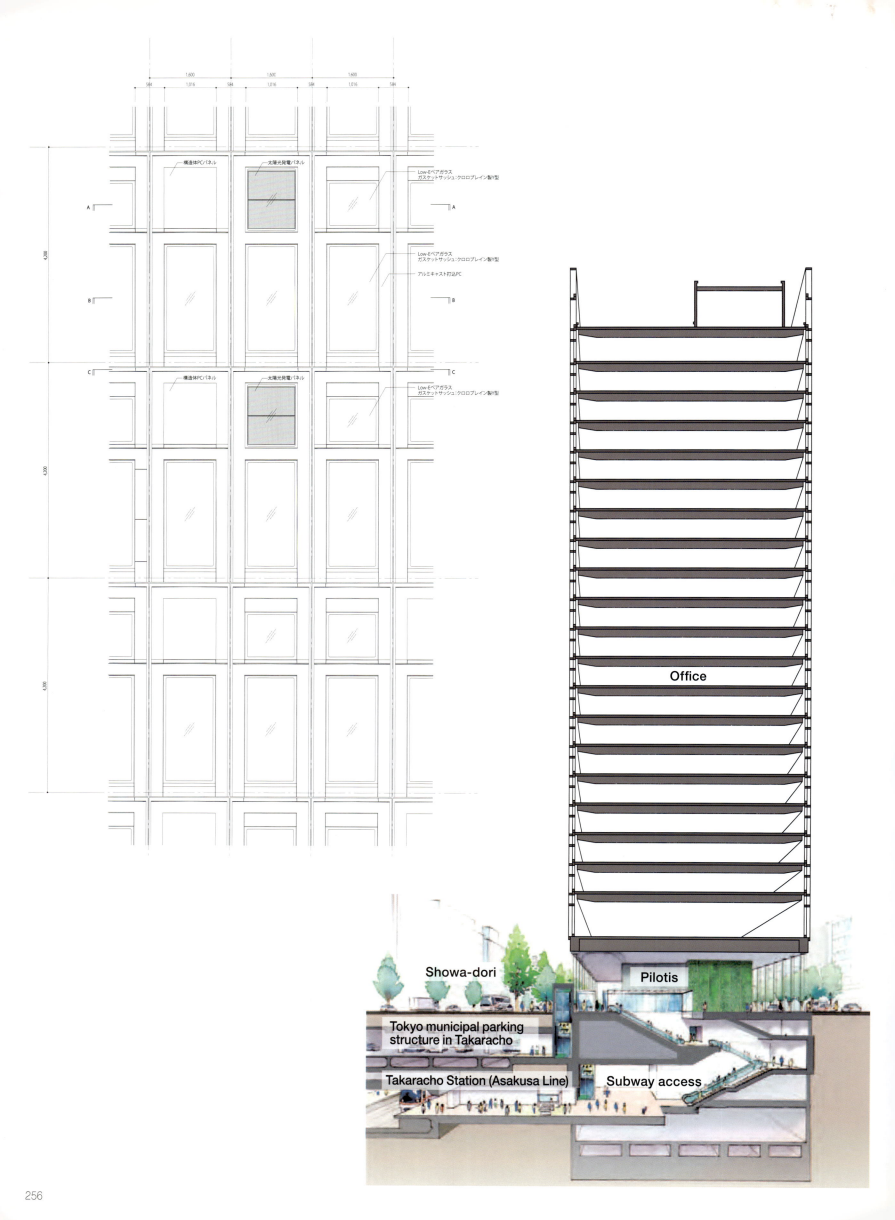

構造体PCパネル　太陽光発電パネル

Low-Eペアガラス
ガスケットサッシュ：クロロプレイン製Y型

Low-Eペアガラス
ガスケットサッシュ：クロロプレイン製Y型
アルミキャスト打込PC

構造体PCパネル　太陽光発電パネル

Low-Eペアガラス
ガスケットサッシュ：クロロプレイン製Y型

A

B

C

1,600　1,600　1,600
584　1,016　584　1,016　584　1,016　584

4,200

4,200

4,200

Office

Showa-dori

Pilotis

Tokyo municipal parking
structure in Takaracho

Takaracho Station (Asakusa Line)

Subway access

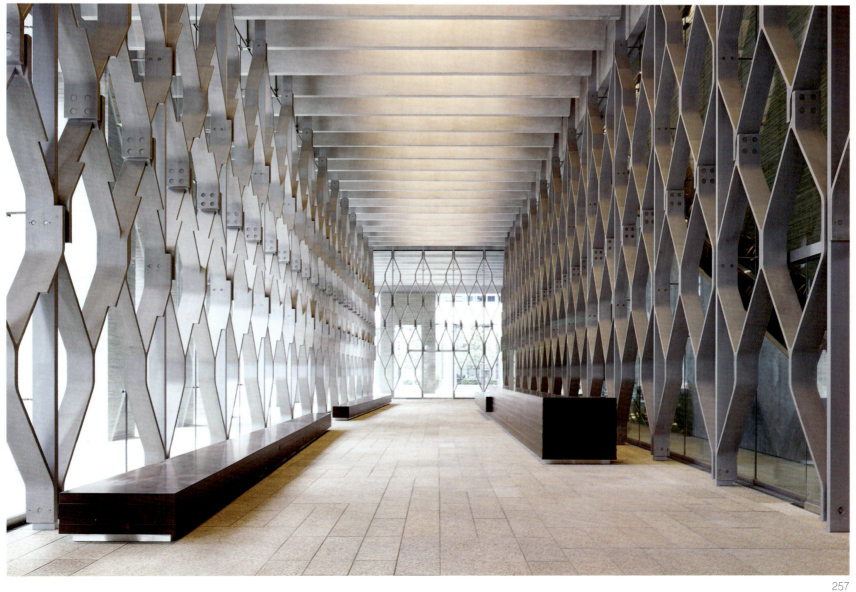

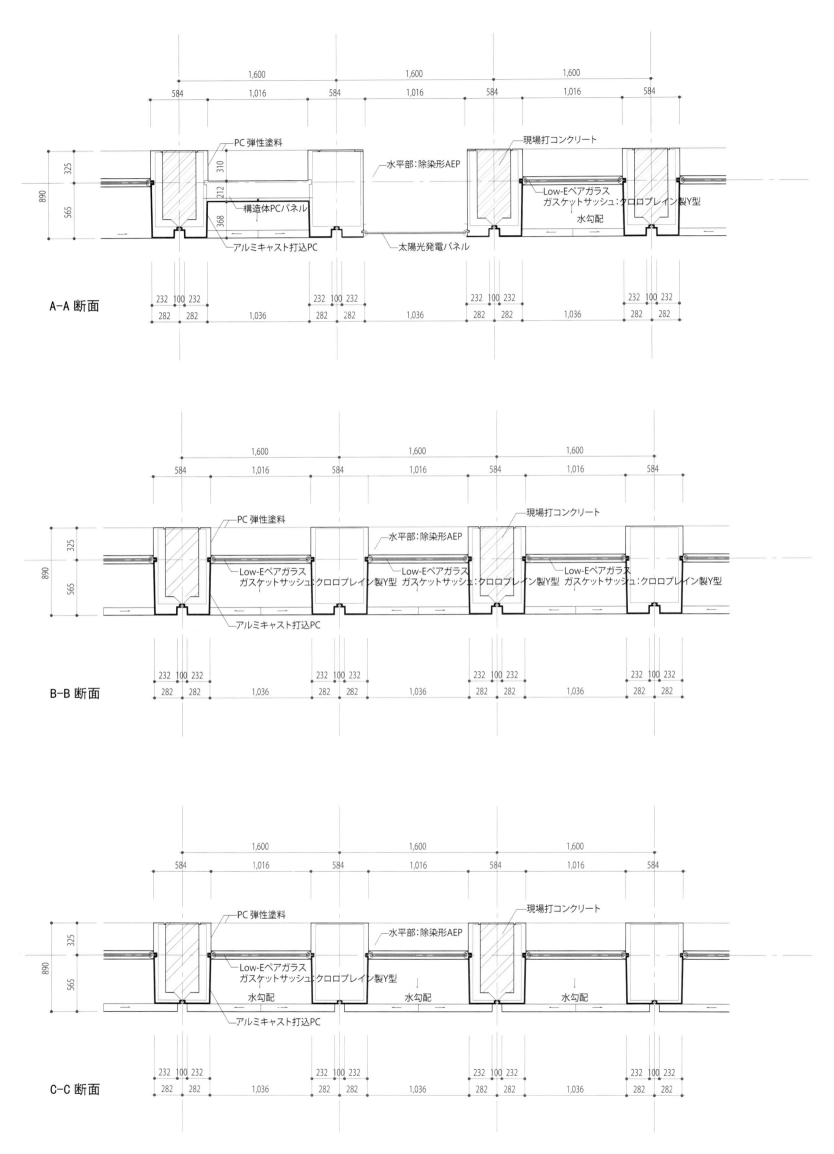

A–A 断面

B–B 断面

C–C 断面

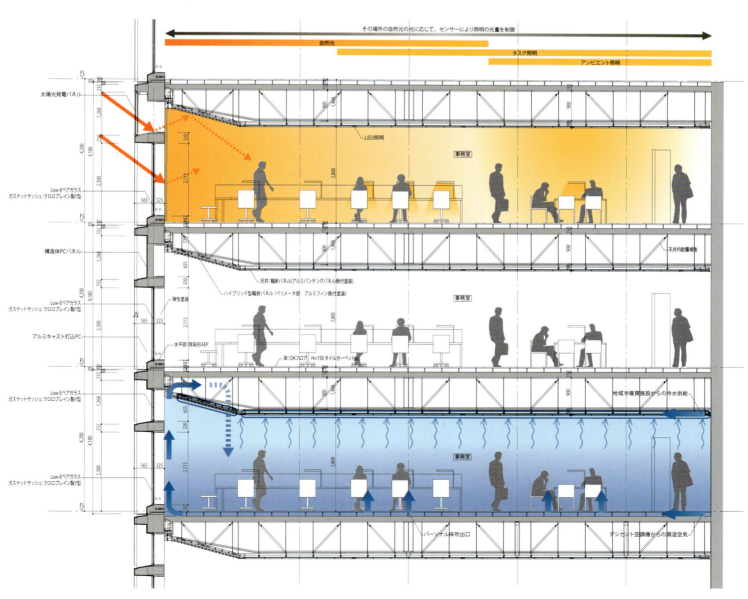

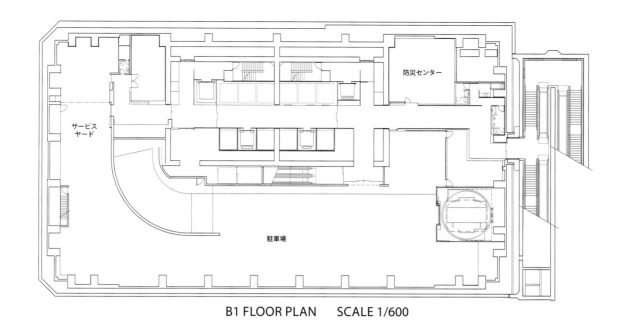

B1 FLOOR PLAN   SCALE 1/600

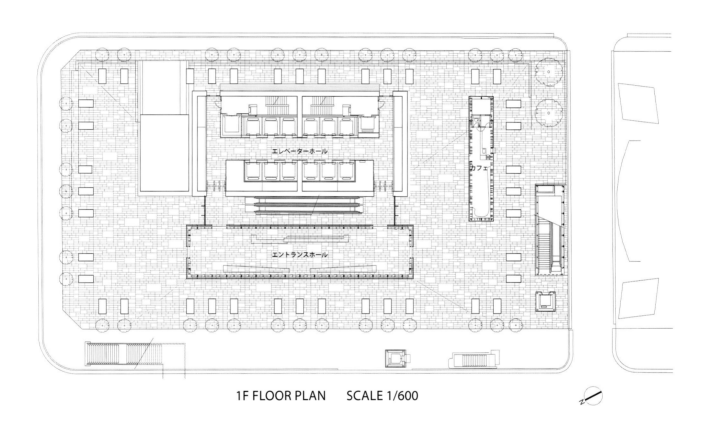

1F FLOOR PLAN   SCALE 1/600

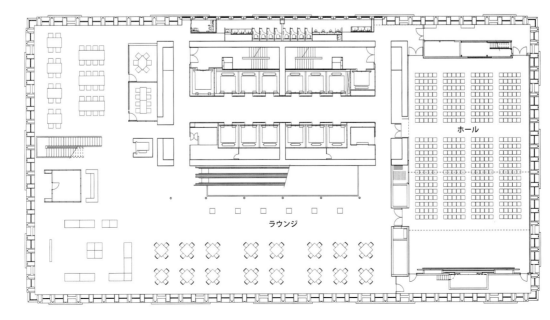

2F FLOOR PLAN   SCALE 1/600

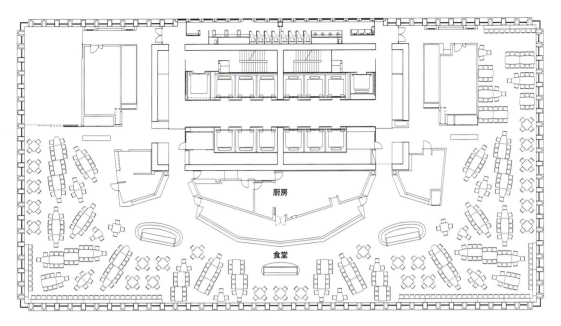

4F FLOOR PLAN　　SCALE 1/600

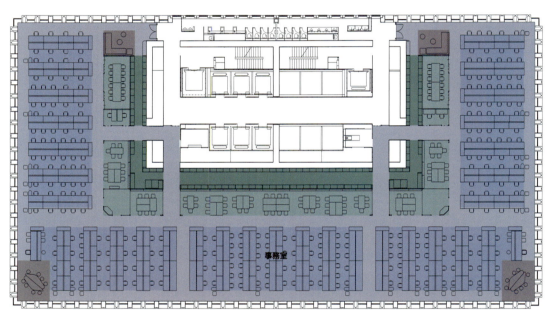

SCALE 1/600

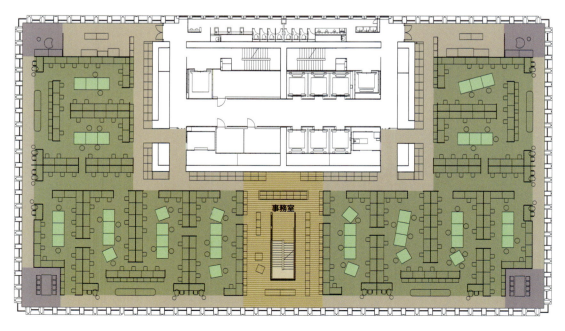

SCALE 1/600

THE LONDON BRIDGE TOWER

伦敦桥大厦

LOCATION: LONDON, UK

ARCHITECT: RENZO PIANO BUILDING WORKSHOP
COLLABORATOR: ADAMSON ASSOCIATES

CLIENT: SELLAR PROPERTY GROUP
PHOTOGRAPHER: ROB TELFORD, PAUL RAFTERY, JOHN SAFA AND RPBW

The London Bridge Tower, which is also known as the Shard, is a 72-storey mixed use tower located besides London Bridge Station on the south bank of the river Thames.

The form of the tower was determined by its prominence on the London skyline. References included the masts of ships docked in the nearby Pool of London and Monet's paintings of the Houses of Parliament.

The slender pyramidal form is suited to the variety of uses proposed: large floor plates for offices at the bottom, public areas and a hotel in the middle, apartments at the top. The final public floors, levels 68~72, accommodate a viewing gallery 240m above street level. Above, the shards continue to 306m. The mix of uses add vibrancy to the project: public access was deemed particularly important for such a significant building in London.

Eight glass shards define the shape and visual quality of the tower. The passive double facade uses low-iron glass throughout, with a mechanized roller blind in the cavity providing solar shading. In the "fractures" between the shards opening vents provide natural ventilation to winter gardens. These can be used as meeting rooms or break-out spaces in the offices and winter gardens on the residential floors. They provide a vital link with the external environment often denied in hermetically sealed buildings.

The main structural element is the slip formed concrete core in the center of the building. It houses the main service risers, lifts and escape stairs. A total of 44 single and double-deck lifts link the key functions with the various entrances at street and station concourse level.

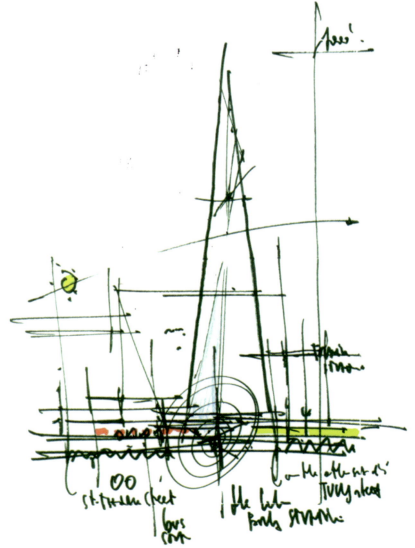

The Spire
Floors 75-95
Height 310m (1,016ft)

The Observatory
Floors 68-72
1,372 sq m (14,768 sq ft)
Height 244m (800ft)

Residences
Floors 53-65
5,720 sq m (61,570 sq ft)
Height 186m (610ft)

Shangri-La Hotel and Spa
Floors 34-52
17,786 sq m (174,889 sq ft)

Bars and Restaurants
Floors 31-33
2,451 sq m (26,382 sq ft)

The Shard Offices
Floors 04-28
55,439 sq m (596,740 sq ft)

The Place Offices
Floors 01~17
39,824 m² (428,664 sq ft)

Roof Terraces

Concourse Level: The Place Office Entrance

Public Piazza

Underground Retail

St Thomas Street: Restaurant Entrance

St Thomas Street: Hotel & Residential Entrance

Concourse Level: Shard Office Entrance

Joiner Street: Underground & Retail Access

伦敦桥大厦，也称为"碎片大厦"，是一座72层的多功能大厦，位于泰晤士河南岸靠近伦敦桥站。

大厦在伦敦天际线的突出部分决定了它的形状，其参照物包括停靠在伦敦附近水域的船舶和莫奈的画作《国会大厦》。

细长的金字塔形状适合大厦的多种用途：底部是用作办公区的大型楼面板，公共区域和酒店位于大厦中部，公寓住宅位于大厦顶部。最后的公共层，第68层到72层，用作观景廊，该观景廊高于街面240米。再往上，大厦高度延伸到306米。多功能用途还增加了项目的活力：在伦敦此类重要的建筑中，公共通道被认为是特别重要的。

八块玻璃片定义了大厦的形状和视觉质量。被动式双层幕墙全部采用超白玻璃，并在凹处安装自动遮光窗帘来遮阳。玻璃片连接处有开口，能为冬景花园提供自然通风。这些地方在办公区可以用作会议室或者分组空间，在住宅区可以用作冬景花园。这些地方是与外部环境相连的重要组成部分，而在密封的建筑中往往是没有的。

大厦的主要结构部件是建筑中心的滑模混凝土芯。它包含了主要服务立管、电梯和疏散楼梯。其共有44部单层和双层电梯，以便连接位于街道和车站大堂的各个入口。

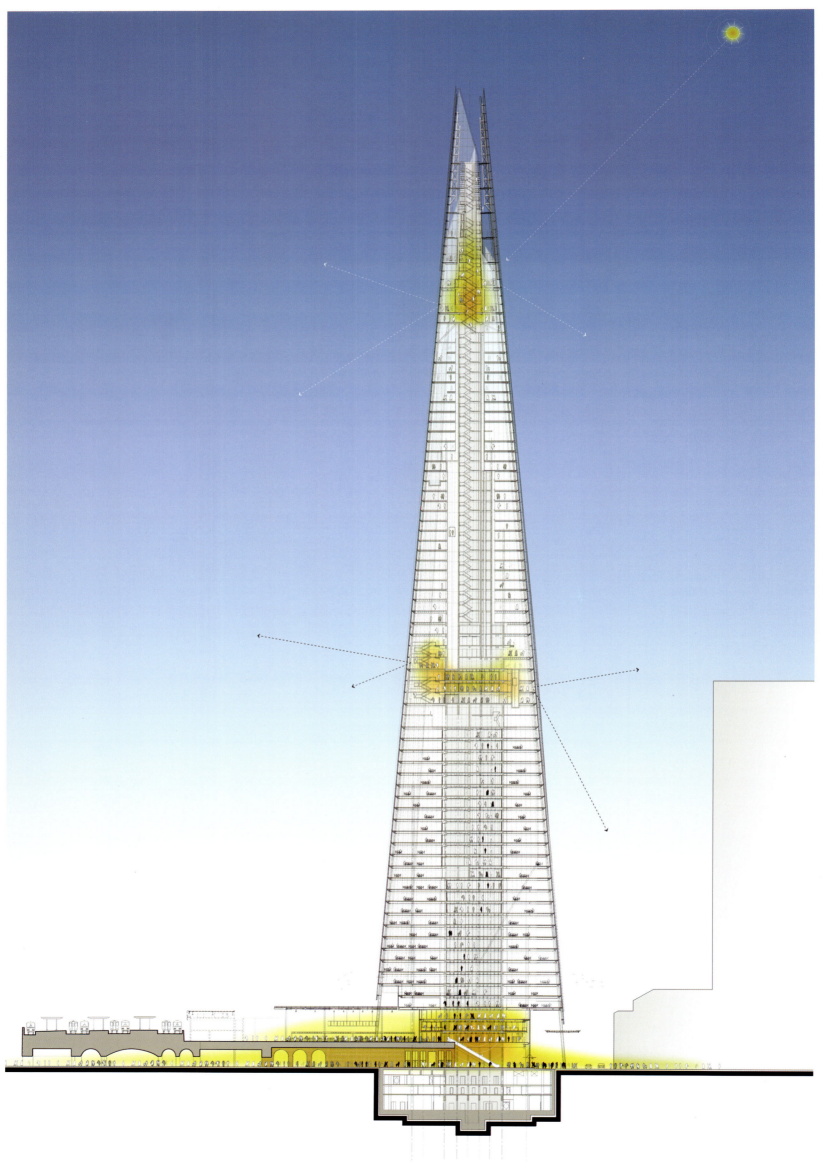

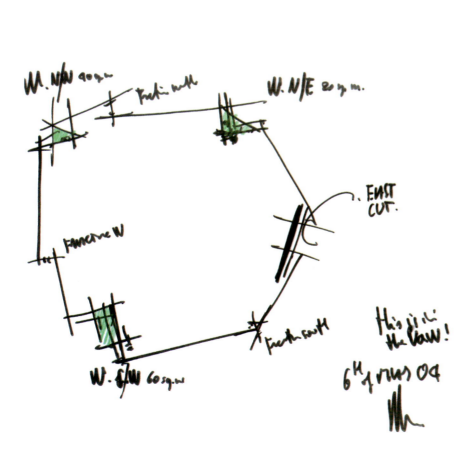

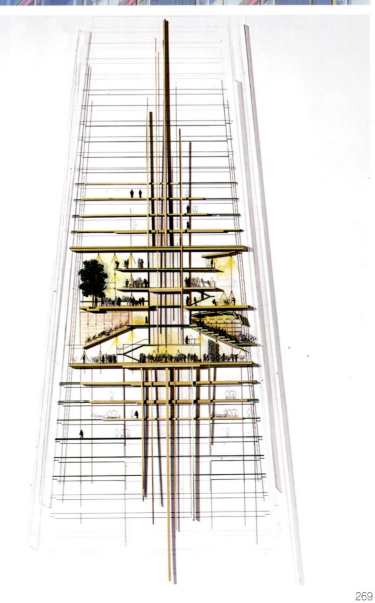

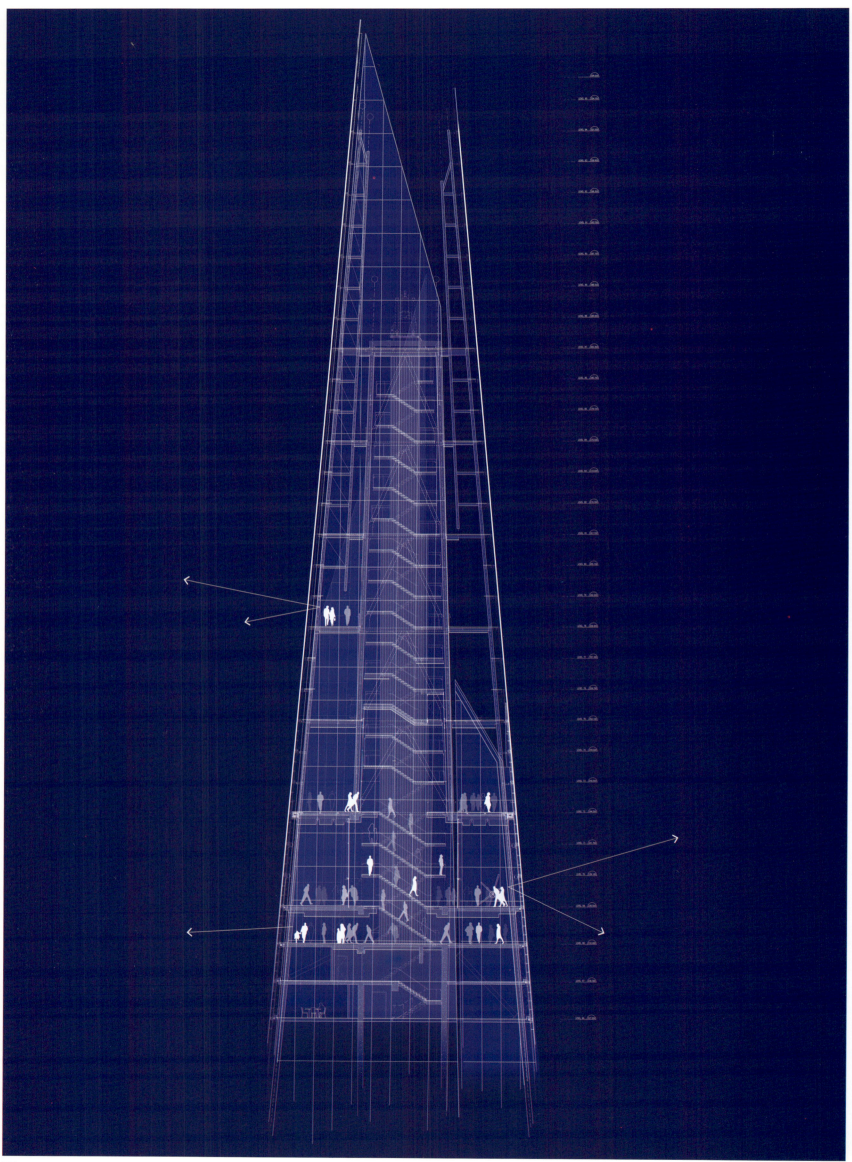

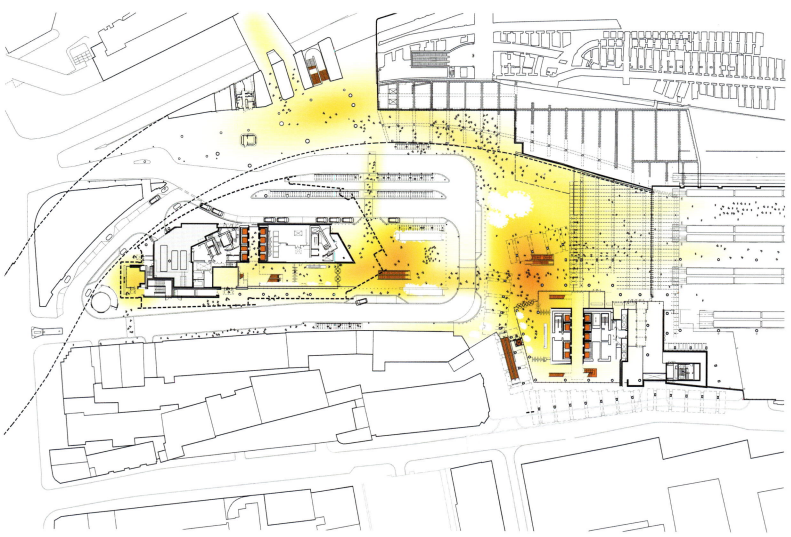

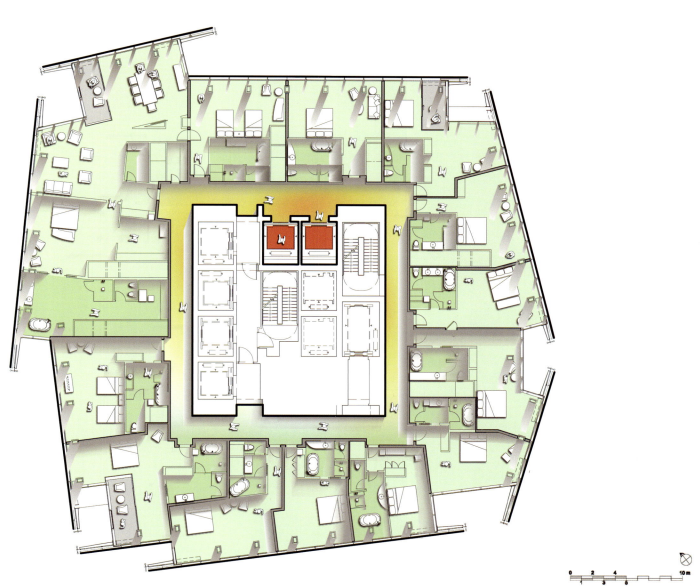

THE BLADE-YONGSAN INTERNATIONAL BUSINESS DISTRICT

龙山国际商务区

LOCATION: SEOUL, KOREA

ARCHITECT: DOMINIQUE PERRAULT ARCHITECTE
ARCHITECT OF THE RECORDS : SAMOO

CLIENT: DREAMHUB-YONGSAN DEVELOPMENT CO., LTD. SEOUL, KOREA.
RENDERING: DPA / ADAGP /LUXIGON

The Yongsan International Business Center, ambitious program of nearly 3 million square meters, is organized as an archipelago of vertical buildings inter-connected a by large park.

By its silhouette and its dynamic allure, the tower establishes itself in the area as a geographical landmark. Its mysterious shape appears like a totem, an iconic figure. It is not a square or a round building, but a rhomboid prism, arranged in a way that makes it look different depending on the angle of approach. Inspired by its slender shape and sharp edges, the tower has been named The Blade.

In the way of a sheath, the skin of the tower is clad with glass, reflecting light and its environment, thus releasing a luminous halo which envelopes the silhouette of the tower. This vibration of the building's skin appears and disappears according

to the viewing angle, creating a living architecture, transforming itself with the movements of the sun and the changes of light.

The project sculptures the void like a luxurious material, offering space, light and views of the grand Seoul landscape. The Grand Lobby, the Business Forum, the Wellness or the Panorama Lobby constitute as many cut-outs in the tower volume, dedicated to promenades and relaxation. This superposition of voids contrasts with the constructed volume of adjacent towers and accentuates the lightness of the tower prism. The voids offer respirations and accommodate collective spaces open to the landscape. At night, they dematerialize the silhouette of the tower, which appears then like a precious carved stone.

OFFICES

EXECUTIVE FLOORS

CHILDCARE

CAFE

LOBBY

RETAIL

TECHNICAL

PARKING

LOADING DOCK

GLASS SUMMIT +292.50 / +311.50

ROOF TOP +279.00 / +298.00
LEVEL L56 +274.50 / +293.50
LEVEL L55 +270.00 / +289.00

LEVEL L54 +261.00 / +280.00
LEVEL L53 +256.50 / +275.50
LEVEL L52 +252.00 / +271.00
LEVEL L51 +247.50 / +266.50
LEVEL L50 +243.00 / +262.00
LEVEL L49 +238.50 / +257.50
LEVEL L48 +234.00 / +253.00
LEVEL L47 +229.50 / +248.50
LEVEL L46 +225.00 / +244.00
LEVEL L45 +220.50 / +239.50
LEVEL L44 +216.00 / +235.00
LEVEL L43 +211.50 / +230.50
LEVEL L42 +207.00 / +226.00
LEVEL L41 +202.50 / +221.50
LEVEL L40 +198.00 / +217.00
LEVEL L39 +193.50 / +212.50
LEVEL L38 +189.00 / +208.00

LEVEL L37 +180.00 / +199.00

LEVEL L36 +171.00 / +190.00
LEVEL L35 +166.50 / +185.50
LEVEL L34 +162.00 / +181.00
LEVEL L33 +157.50 / +176.50
LEVEL L32 +153.00 / +172.00
LEVEL L31 +148.50 / +167.50
LEVEL L30 +144.00 / +163.00
LEVEL L29 +139.50 / +158.50
LEVEL L28 +135.00 / +154.00
LEVEL L27 +130.50 / +149.50
LEVEL L26 +126.00 / +145.00
LEVEL L25 +121.50 / +140.50
LEVEL L24 +117.00 / +136.00
LEVEL L23 +112.50 / +131.50
LEVEL L22 +108.00 / +127.00
LEVEL L21 +103.50 / +122.50
LEVEL L20 +99.00 / +118.00
LEVEL L19 +94.50 / +113.50

LEVEL L18 +85.50 / +104.50

LEVEL L17 +78.50 / +95.50
LEVEL L16 +72.00 / +91.00
LEVEL L15 +67.50 / +86.50
LEVEL L14 +63.00 / +82.00
LEVEL L13 +58.50 / +77.50
LEVEL L12 +54.00 / +73.00
LEVEL L11 +49.50 / +68.50
LEVEL L10 +45.00 / +64.00
LEVEL L9 +40.50 / +59.50
LEVEL L8 +36.00 / +55.00
LEVEL L7 +31.50 / +50.50
LEVEL L6 +27.00 / +46.00
LEVEL L5 +22.50 / +41.50
LEVEL L4 +18.00 / +37.00
LEVEL L3 +13.50 / +32.50
LEVEL L2 +9.00 / +28.00
LEVEL L1 ±0.00 / +19.00

LEVEL B1 -12.00 / +7.00
LEVEL B2 -19.00 / +00.00
LEVEL B3 -23.00 / +4.00
LEVEL B4 -4.00 / -27.00
LEVEL B5 -6.00 / -31.00
LEVEL B6 -12.00 / -35.00 / -16.00
LEVEL B7- B8 -43.00 / -34.00

HELICOPTER RESCUE AREA

PANORAMA LOBBY
PUBLIC LOBBY

TECHNICAL

EXECUTIVE FLOORS

HIGH RISE
14 OFFICE FLOORS

WELLNESS LOBBY
HIGH RISE SKY LOBBY

TECHNICAL

REFUGE + TECHNICAL

MID RISE
15 OFFICE FLOORS

BUSINESS FORUM
LOW RISE SKY LOBBY

TECHNICAL

REFUGE + TECHNICAL

LOW RISE
15 OFFICE FLOORS

CHILDCARE
CAFE
LE GRAND LOBBY
ART SPACE
URBAN LIFE

RETAIL

PARKING

TECNICAL ROOMS

EXECUTIVE LOUNGE

WORLD GOURMET

SHOWROOMS

LOADING DECK

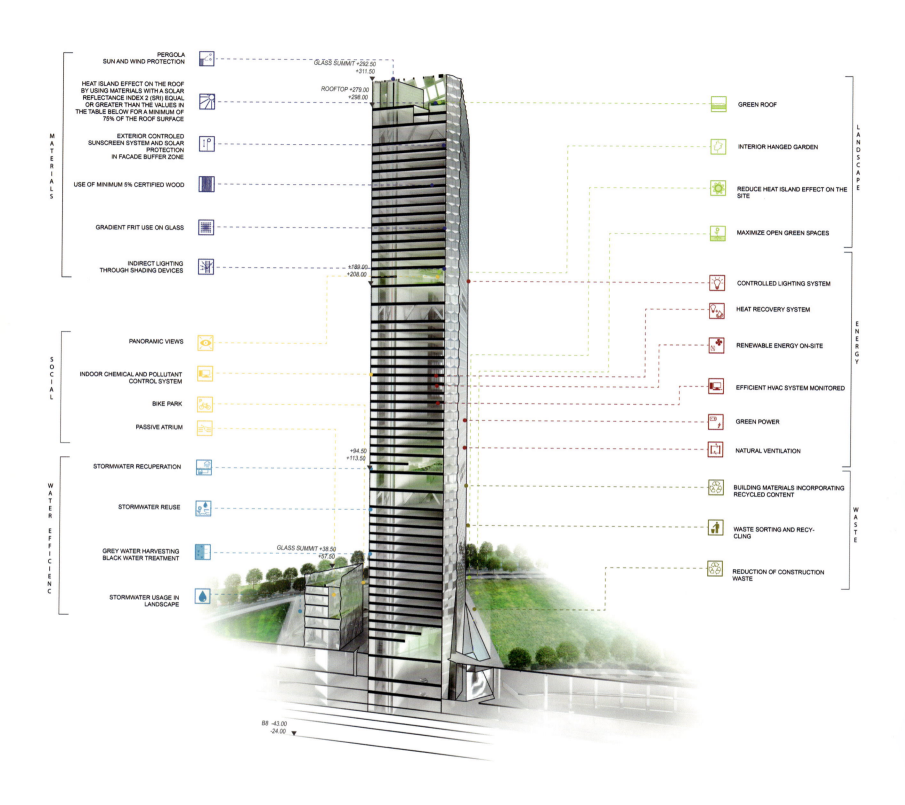

PERGOLA
SUN AND WIND PROTECTION

HEAT ISLAND EFFECT ON THE ROOF
BY USING MATERIALS WITH A SOLAR
REFLECTANCE INDEX 2 (SRI) EQUAL
OR GREATER THAN THE VALUES IN
THE TABLE BELOW FOR A MINIMUM OF
75% OF THE ROOF SURFACE

EXTERIOR CONTROLED
SUNSCREEN SYSTEM AND SOLAR
PROTECTION
IN FACADE BUFFER ZONE

USE OF MINIMUM 5% CERTIFIED WOOD

GRADIENT FRIT USE ON GLASS

INDIRECT LIGHTING
THROUGH SHADING DEVICES

MATERIALS

PANORAMIC VIEWS

INDOOR CHEMICAL AND POLLUTANT
CONTROL SYSTEM

BIKE PARK

PASSIVE ATRIUM

SOCIAL

STORMWATER RECUPERATION

STORMWATER REUSE

GREY WATER HARVESTING
BLACK WATER TREATMENT

STORMWATER USAGE IN
LANDSCAPE

WATER EFFICIENC

GLASS SUMMIT +292.50
+311.50

ROOFTOP +279.00
+298.00

+189.00
+208.00

+94.50
+113.50

GLASS SUMMIT +38.50
+57.50

B8 -43.00
-24.00

GREEN ROOF

INTERIOR HANGED GARDEN

REDUCE HEAT ISLAND EFFECT ON THE
SITE

MAXIMIZE OPEN GREEN SPACES

LANDSCAPE

CONTROLLED LIGHTING SYSTEM

HEAT RECOVERY SYSTEM

RENEWABLE ENERGY ON-SITE

EFFICIENT HVAC SYSTEM MONITORED

GREEN POWER

NATURAL VENTILATION

ENERGY

BUILDING MATERIALS INCORPORATING
RECYCLED CONTENT

WASTE SORTING AND RECY-
CLING

REDUCTION OF CONSTRUCTION
WASTE

WASTE

龙山国际商务中心，一个近300万平方米的雄心勃勃的项目，形成了一个高楼林立的群岛，各楼通过一个大型广场相互连接。

由于其轮廓及动感魅力，塔楼已成为该地区一个具有里程碑意义的地标。其不可思议的外形看上去好像一个图腾、一个偶像人物。塔楼不是正方形或圆形的建筑，而是一个包含菱形的棱柱体，从不同的角度看会给人不同的感觉。由于其细长的形状和锐利的边缘，塔楼已被命名为"刀锋"。

关于护层，塔楼外表覆盖着玻璃，通过反射光线和周围环境，从而环绕大厦形成了一个

明亮、美丽的光环。随着视角的不同，大厦外表给人的这种光线感应或隐或现，形成了一个有生命的建筑，并随着太阳的移动和光线的变化而不停地变换。

项目还雕刻出很多的空洞空间，这些空间闪闪发亮，便于提供空间、光线和欣赏首尔的壮美景观。宏伟的大厅、商业会议区、健身区和全景大厅都是挖空空洞的形式，能使人们得到休闲放松。这种空洞空间的叠加与邻近大楼的建筑方式形成了鲜明的对比，并突出了棱柱体塔楼的亮度。空洞空间还提供了呼吸空隙和欣赏开放景观的集体空间。在晚上，空洞空间消失在塔楼的轮廓里，看上去好像珍贵的石雕一样。

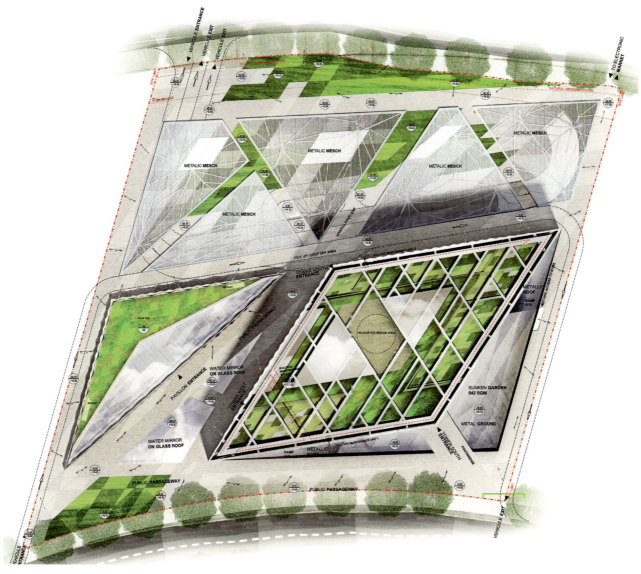

PULSE TOWER

涌动大厦

LOCATION: MONTERREY, MEXICO

ARCHITECT: ROJKIND ARQUITECTOS
FOUNDING COLLABORATOR: MICHEL ROJKIND

CLIENT: ORANGE INVESTMENTS
AREA: 80,000 m²

The building is positioned within the terrain, taking into account the two circumstances which determine it: the surrounding buildings and the need to weave a harmonious urban profile. The building's mass is a transition between neighboring heights, beginning as an 11-floor volume on its north side, and becoming a 28-floor tower on the south. Between these two volumes a plaza emerges. Its embracing gesture is open to the city as an invitation. The bold, innovative commercial plaza welcomes visitors with open halls, drawing them through the plaza into an open air market. Terraces and platforms transition between the street and commercial zone making the relationship with the space more intimate at each level. Stripes of vegetation, a variety of textures, and water emerging from the building dispel the notion of an arid street.

The office and residential lobbies are located along paths through the plaza. The corporate office lobbies are located on the low, north side of the plaza while the residential lobby and the instant offices are located more privately, near the plaza's peak.

Two parking levels can be found below the plaza, the rest are located behind the main spaces at the building's facade.

The building's skin is borne from the plaza, as if it were a geological mantle thrust upwards from a plateau. Additionally, the skin is perforated with a system of horizontal windows that have no apparent order. This creates the ephemeral effect of a coded message, which is constantly encrypted anew as the day passes. At night this system is enriched by LED panels interlaid between windows, which pulse and change color, giving the sensation of a breathing building.

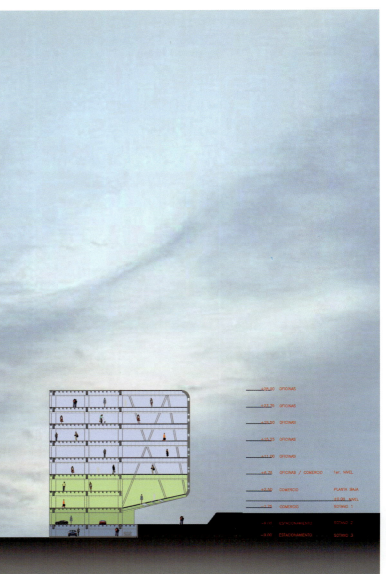

AMENITIES +19.50
ESTACIONAMIENTO VIVIENDA +18.00
ESTACIONAMIENTO VIVIENDA +12.00
ESTACIONAMIENTO VIVIENDA +9.00
ESTACIONAMIENTO OFICINAS +6.00
ESTACIONAMIENTO OFICINAS +3.00
ESTACIONAMIENTO OFICINAS ±0.00 NIVEL
ESTACIONAMIENTO COMERCIO -3.00
ESTACIONAMIENTO COMERCIO -6.00
ESTACIONAMIENTO OFICINAS -9.00
ESTACIONAMIENTO OFICINAS -12.00

+33.00 OFICINAS
+28.00 OFICINAS
+23.75 OFICINAS
+19.50 OFICINAS
+13.25 OFICINAS
+11.00 OFICINAS
+8.25 OFICINAS/COMERCIO   1er. NIVEL
+3.50 COMERCIO   PLANTA BAJA
-1.75 COMERCIO   SOTANO 1
-4.50 ESTACIONAMIENTO COMERCIO   SOTANO 2
-9.00 ESTACIONAMIENTO OFICINAS   SOTANO 3

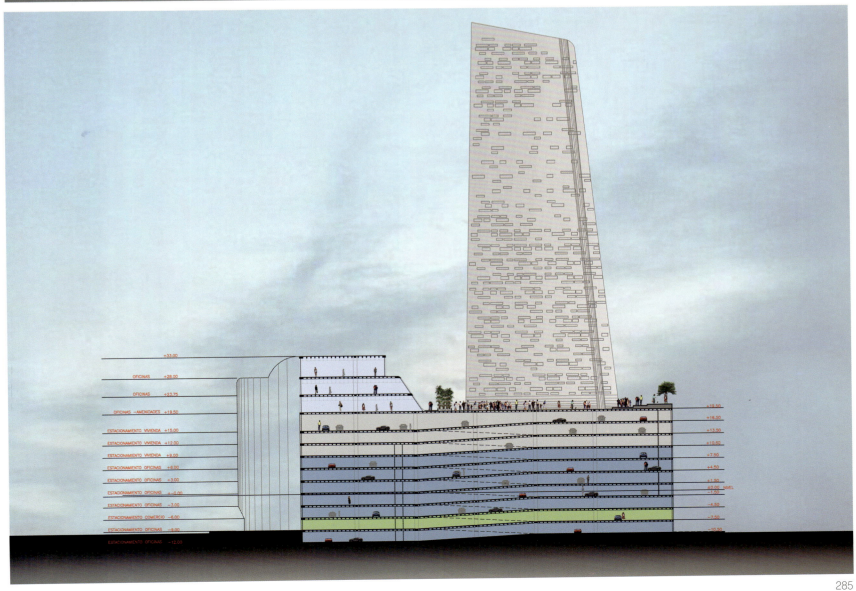

+33.00
OFICINAS +28.00
+23.75
OFICINAS -AMENIDADES +18.50
ESTACIONAMIENTO VIVIENDA +15.00
ESTACIONAMIENTO VIVIENDA +12.00
ESTACIONAMIENTO VIVIENDA +9.00
ESTACIONAMIENTO OFICINAS +6.00
ESTACIONAMIENTO OFICINAS +3.00
ESTACIONAMIENTO OFICINAS ±0.00
ESTACIONAMIENTO OFICINAS -3.00
ESTACIONAMIENTO COMERCIO -6.00
ESTACIONAMIENTO OFICINAS -9.00
ESTACIONAMIENTO OFICINAS -12.00

+18.50
+16.50
+13.50
+10.50
+7.50
+4.50
+1.50
±0.00 NIVEL
-7.50
-4.50
-7.50
-10.50

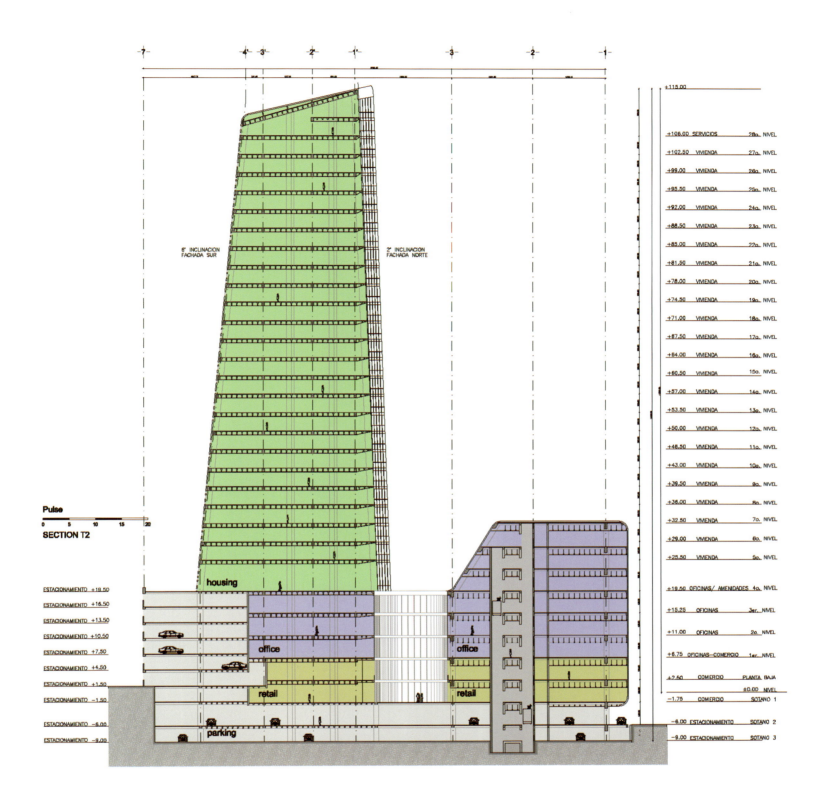

Pulse

0 5 10 15 20

SECTION T2

+115.00

+106.00 SERVICIOS   28a. NIVEL
+102.50 VIVIENDA   27a. NIVEL
+99.00   VIVIENDA   26a. NIVEL
+95.50   VIVIENDA   25a. NIVEL
+92.00   VIVIENDA   24a. NIVEL
+88.50   VIVIENDA   23a. NIVEL
+85.00   VIVIENDA   22a. NIVEL
+81.50   VIVIENDA   21a. NIVEL
+78.00   VIVIENDA   20a. NIVEL
+74.50   VIVIENDA   19a. NIVEL
+71.00   VIVIENDA   18a. NIVEL
+67.50   VIVIENDA   17a. NIVEL
+64.00   VIVIENDA   16a. NIVEL
+60.50   VIVIENDA   15a. NIVEL
+57.00   VIVIENDA   14a. NIVEL
+53.50   VIVIENDA   13a. NIVEL
+50.00   VIVIENDA   12a. NIVEL
+46.50   VIVIENDA   11a. NIVEL
+43.00   VIVIENDA   10a. NIVEL
+39.50   VIVIENDA   9a. NIVEL
+36.00   VIVIENDA   8a. NIVEL
+32.50   VIVIENDA   7a. NIVEL
+29.00   VIVIENDA   6a. NIVEL
+25.50   VIVIENDA   5a. NIVEL

+19.50   OFICINAS / AMENIDADES   4a. NIVEL
+15.25   OFICINAS   3er. NIVEL
+11.00   OFICINAS   2a. NIVEL
+6.75   OFICINAS—COMERCIO   1er. NIVEL
+2.60   COMERCIO   PLANTA BAJA
±0.00   NIVEL
-1.75   COMERCIO   SOTANO 1
-6.00   ESTACIONAMIENTO   SOTANO 2
-9.00   ESTACIONAMIENTO   SOTANO 3

6° INCLINACION FACHADA SUR

2° INCLINACION FACHADA NORTE

housing
office   office
retail   retail
parking

ESTACIONAMIENTO +19.50
ESTACIONAMIENTO +16.50
ESTACIONAMIENTO +13.50
ESTACIONAMIENTO +10.50
ESTACIONAMIENTO +7.50
ESTACIONAMIENTO +4.50
ESTACIONAMIENTO +1.50
ESTACIONAMIENTO -1.50
ESTACIONAMIENTO -6.00
ESTACIONAMIENTO -9.00

大厦的位置考虑到了两个决定性因素：周边建筑，以及构建和谐城市风貌的需求。大厦的北楼高 11 层，南楼 28 层，中间是一个购物中心，高度由北向南上升，与周边建筑的高度相连。这种弧形的高度，就像城市的一张邀请函，吸引着各界人士。标新立异的商业购物中心用开放的大厅欢迎着各界人士。穿过购物中心，是一个露天集市。在街道和商业区中间有一些过渡的门廊与平台，让每一层的空间关系更加亲密。层层的植被、多变的质感、从大厦里流出的水驱散了街道给人的干燥之感。

在通往购物中心的小路两边，是一些办公楼和住宅楼的入口。公司办公区的大厅入口在购物中心北边较低的楼层，而住宅区的大厅和即时出租办公室则位于购物中心顶端更加私密的位置。

在购物中心下面有两层停车场，其他停车场则位于大厦正楼的后面。

大厦的外立面设计以购物中心为核心向两边延伸，就像从一个高原向外延伸凸出的地质逆断层一样。此外，水平方向上的众多窗户还构成了外立面上的无数孔眼，这些孔眼的排列没有明显的规则，似乎瞬间就形成了一个编码信息，并且随着光线的变化而不断发生变化。到了晚上，窗户间的 LED 照明板更加丰富了外立面的层次，色彩涌动，变幻万千，仿佛大厦能够呼吸一般。

ACCESS FLOOR PLAN

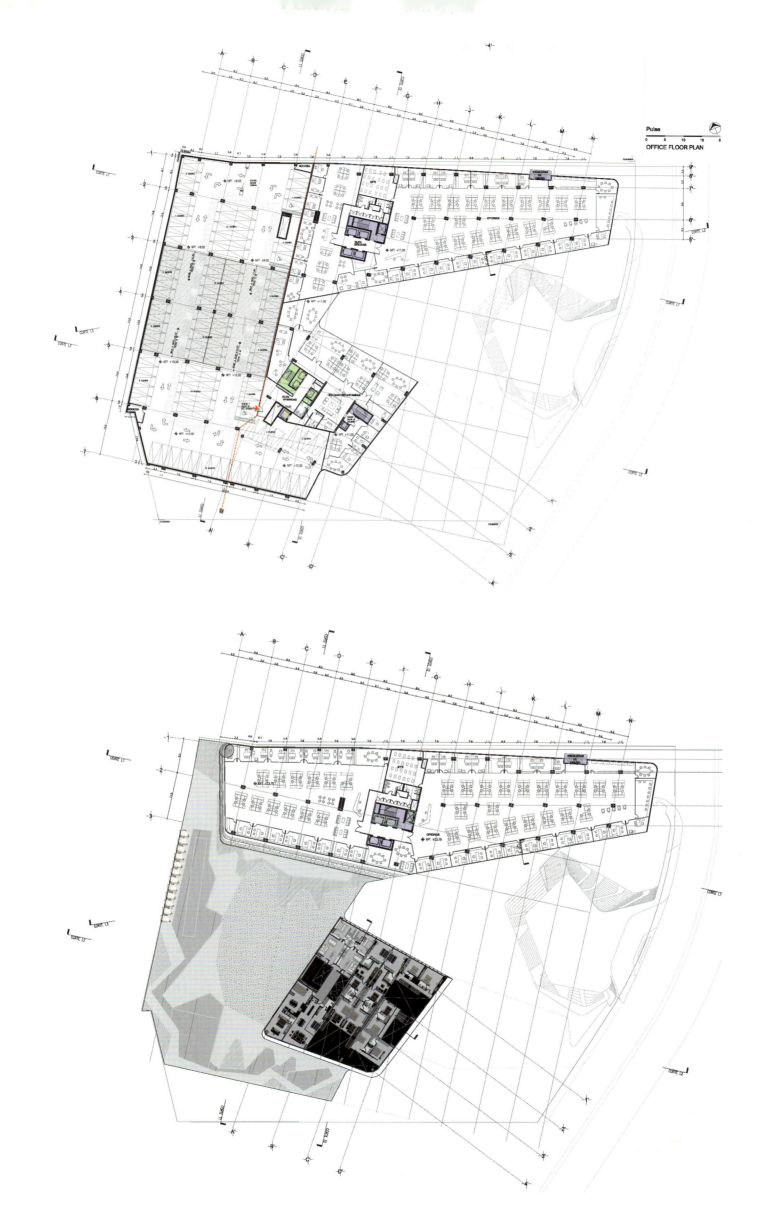

GUANGFA SECURITIES HEADQUARTERS

广发证券总部大楼

LOCATION:
GUANGZHOU CITY, GUANGDONG PROVINCE, CHINA

ARCHITECT: JOHANNES JAEGER

FIRM: JAEGER AND PARTNER ARCHITECTS

AREA: 150,000 m²

Guangfa Securities Headquarters is a high-rise grade-A office building in CBD central area, and a new landmark for Zhujiang New Town. It is mainly used to serve high-end subscribers of GF securities, finance, investment banks, and other related financial products.

The office building consists of two zones. The upper part accomadates the headquarters offices of Guangfa Securities, while the lower portion is rented to other financial companies. The design generates an impression of a building inside the building. An internal and distinctive sky lobby is created at the upper zone. The whole design combines structure with form in an elegant way. The building incorporates solar active layers on the glazing for electric energy, which meets the high requirements of modern offices.

At such an important site will stand an outstanding landmark for Guangzhou: the Headquarters of GF Securities, which integrates elegance and strength, successfully taking the challenges of various high-standard needs of modern offices.

It is distinguished with its international standard design, its respect to the surrounding area and the urban fabric, its high-efficient master plan, and its unique transition space in podium building. The headquarters expresses the concept of combining interior space with facade, creating a symbolic building for the pride of the transitional city and its citizens.

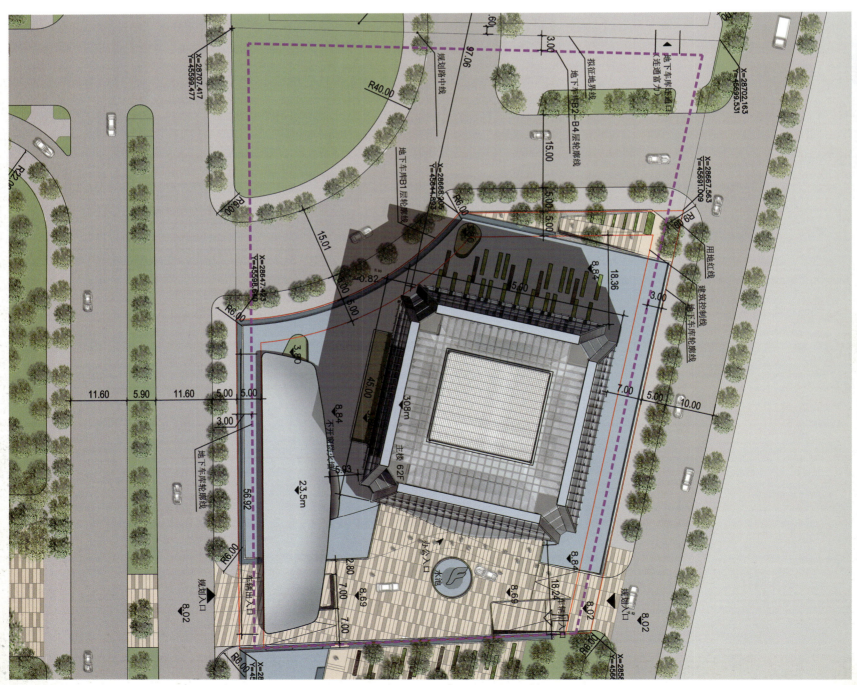

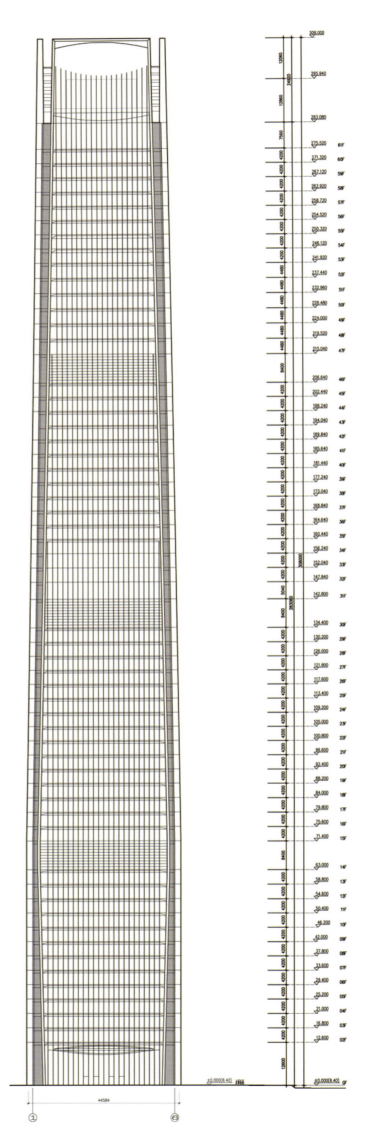

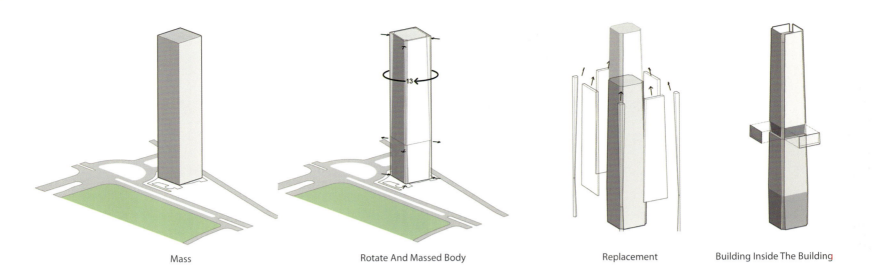

Mass

Rotate And Massed Body

Replacement

Building Inside The Building

Rental Space Life

Guangfa Space Life

Escape Routes

Guangfa Lowrise

Guangfa Highrise

Guangfa Express Life

Rental Lowrise

Rental Highrise

CORE ANALYSIS

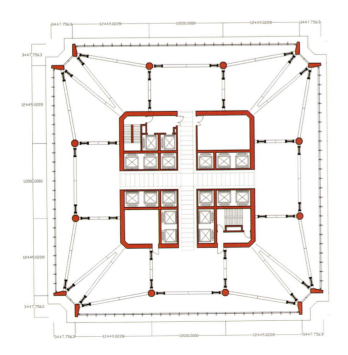

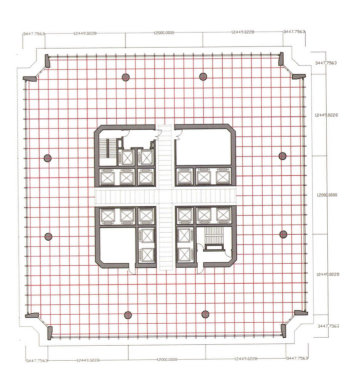

广发证券总部大楼位于珠江新城 CBD 中心区，是一座甲级超高层办公大楼。主要服务于广发证券、金融、投资银行及其他金融产业链相关的高端用户。建筑建成后将成为珠江新城的一座新地标。

办公大楼主要分为两个区域。上部区域为广发证券总部自用区，而下部则租用给其他金融行业公司。整体建筑具有楼中楼的印象，总部自用部分也有其独特的空中大堂。广发证券总部大楼形体优雅，结构与外形设计融为一体，引入太阳能光伏发电玻璃，并满足了现代办公环境的各种高要求。

在这个非常特别的地理位置上，广州将迎来一座杰出的重要的地标性建筑——广发证券总部大厦。一座融合了优雅外表与有力形状，并综合地满足了现代工作环境对建筑的各种复杂而先进的要求。

以国际标准设计、尊重环境、尊重城市设计、高效的平面，和具有特质的裙楼灰空间，使得广发证券总部大楼与众不同。该总部大楼完美地将室内与外观表达设计结合起来，创造出一个转型的城市和令市民引以为荣的标志性建筑。

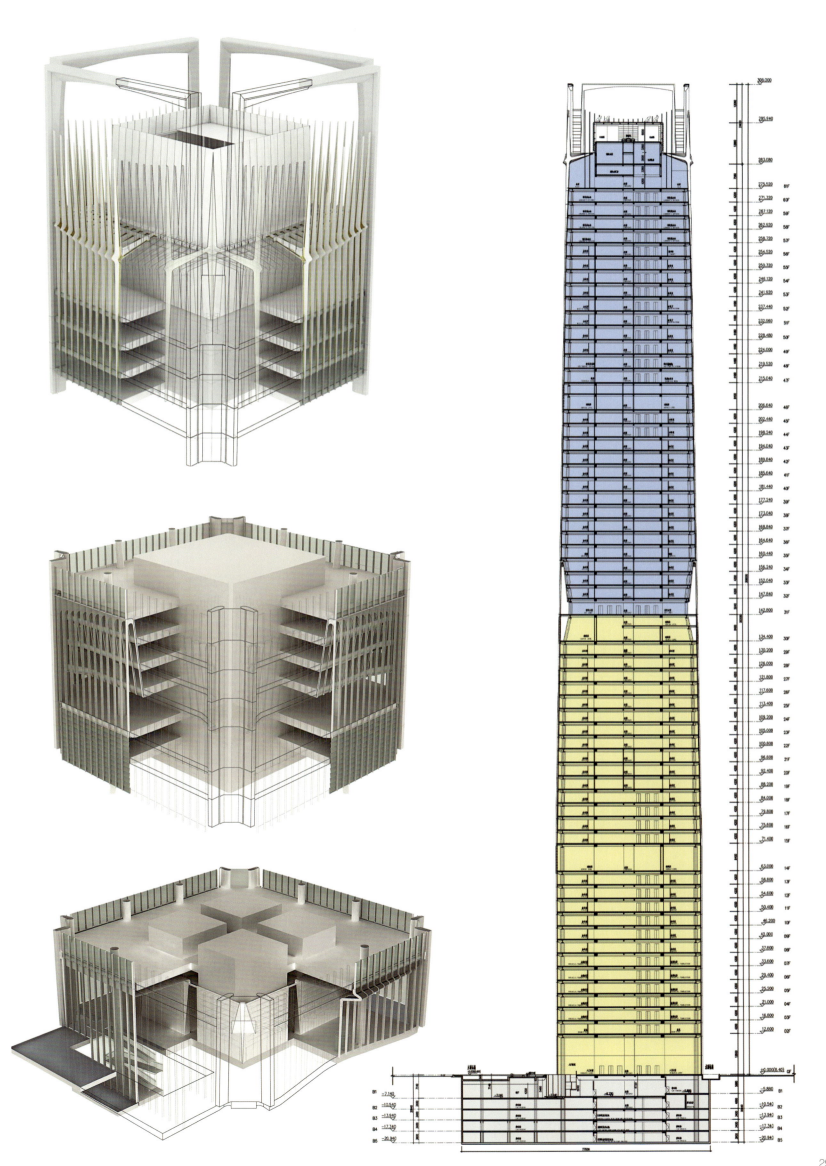

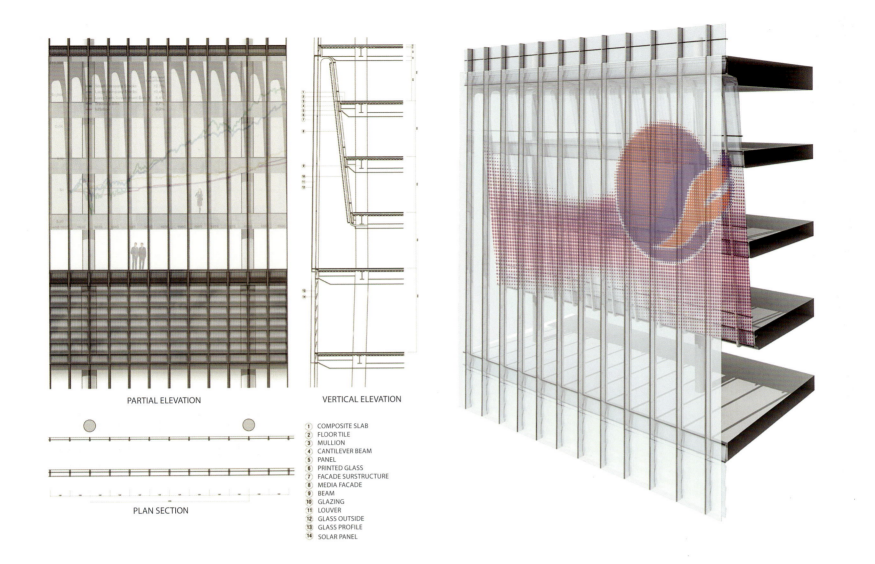

PARTIAL ELEVATION

VERTICAL ELEVATION

PLAN SECTION

1. COMPOSITE SLAB
2. FLOOR TILE
3. MULLION
4. CANTILEVER BEAM
5. PANEL
6. PRINTED GLASS
7. FACADE SURSTRUCTURE
8. MEDIA FACADE
9. BEAM
10. GLAZING
11. LOUVER
12. GLASS OUTSIDE
13. GLASS PROFILE
14. SOLAR PANEL

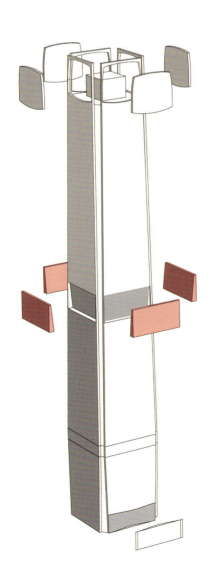

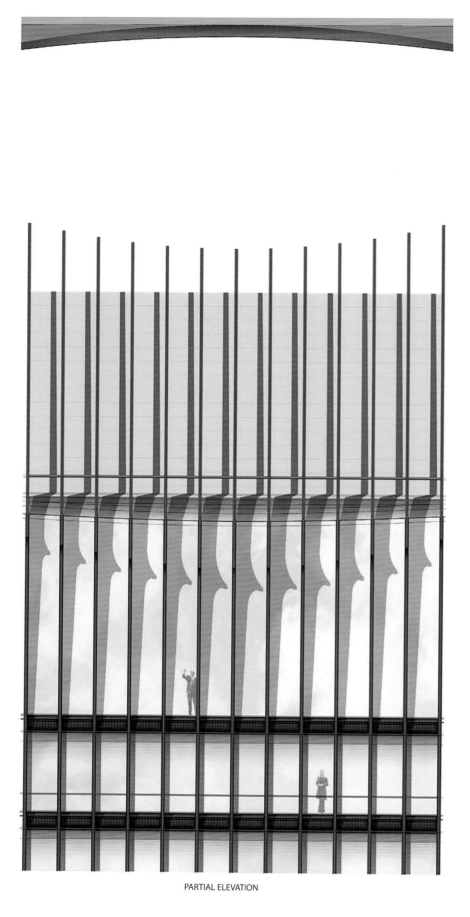

PARTIAL ELEVATION

VERTICAL ELEVATION

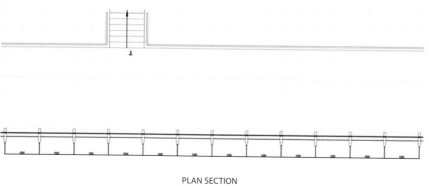

PLAN SECTION

① GLASS COVERED STEEL BEAM
② FACADE SUBSTRUCTURE
③ BRACING
④ CROSS BEAM
⑤ GLAZING
⑥ ELEVATED FLOOR
⑦ COLUMN
⑧ LAMELLA CLADDING
⑨ COLUMN
⑩ CROSS BEAM
⑪ SURSTRUCTURE
⑫ FACADE ANCHORING
⑬ INTERIOR CLADDING

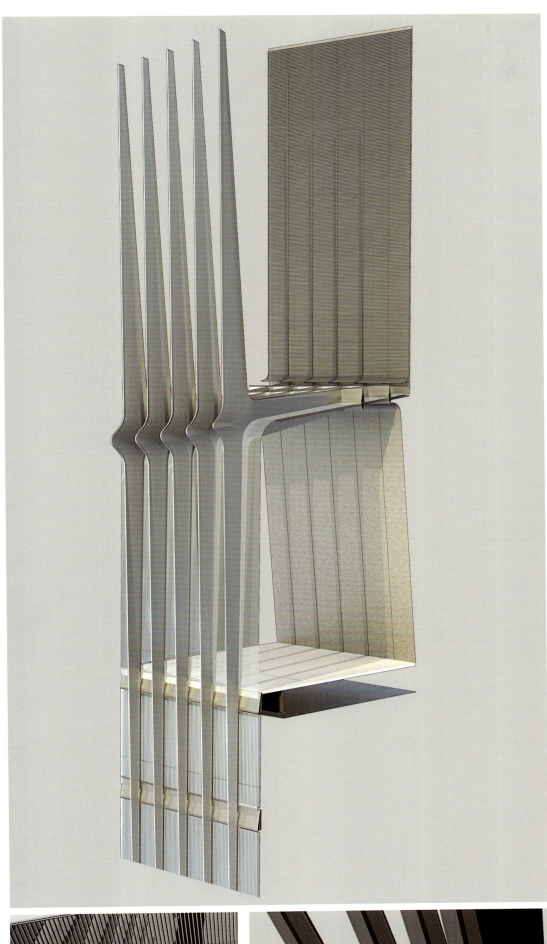

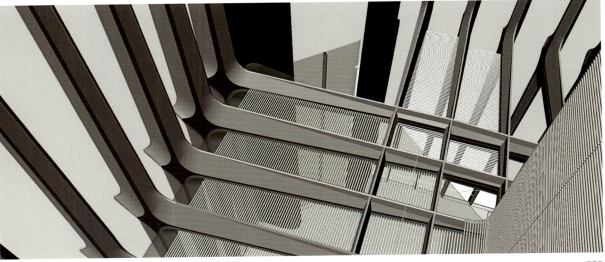

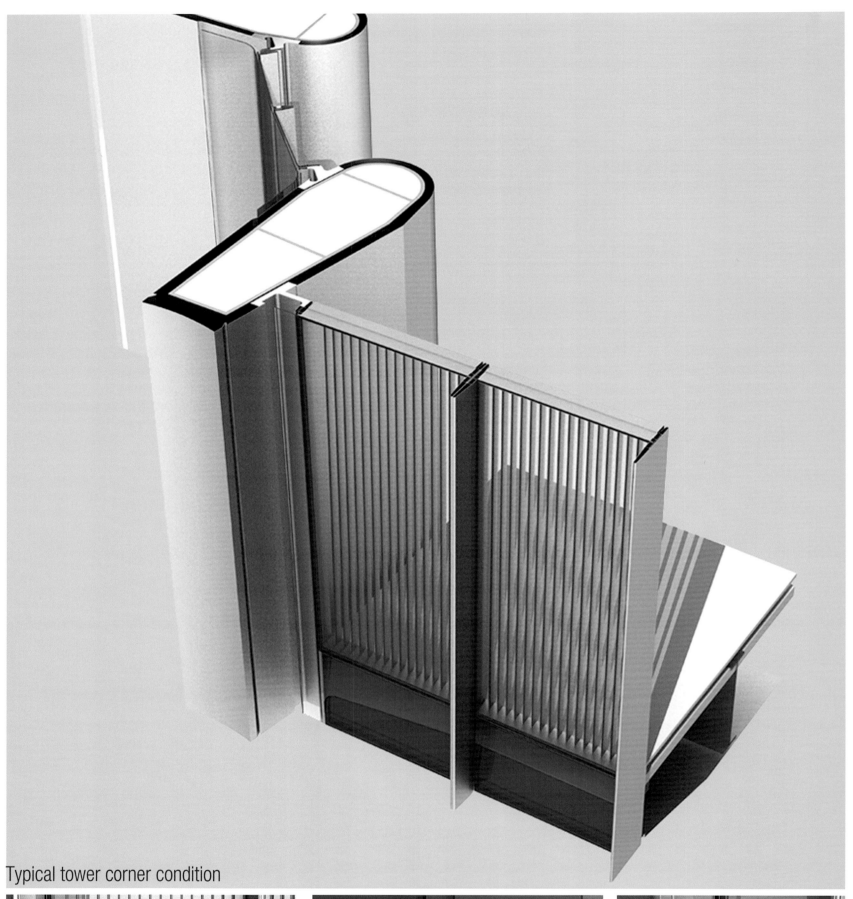

Typical tower corner condition

Detail of spandrel glass extension

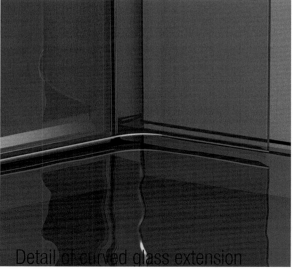

Detail of curved glass extension

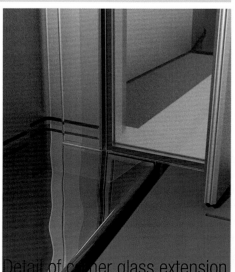

Detail of corner glass extension

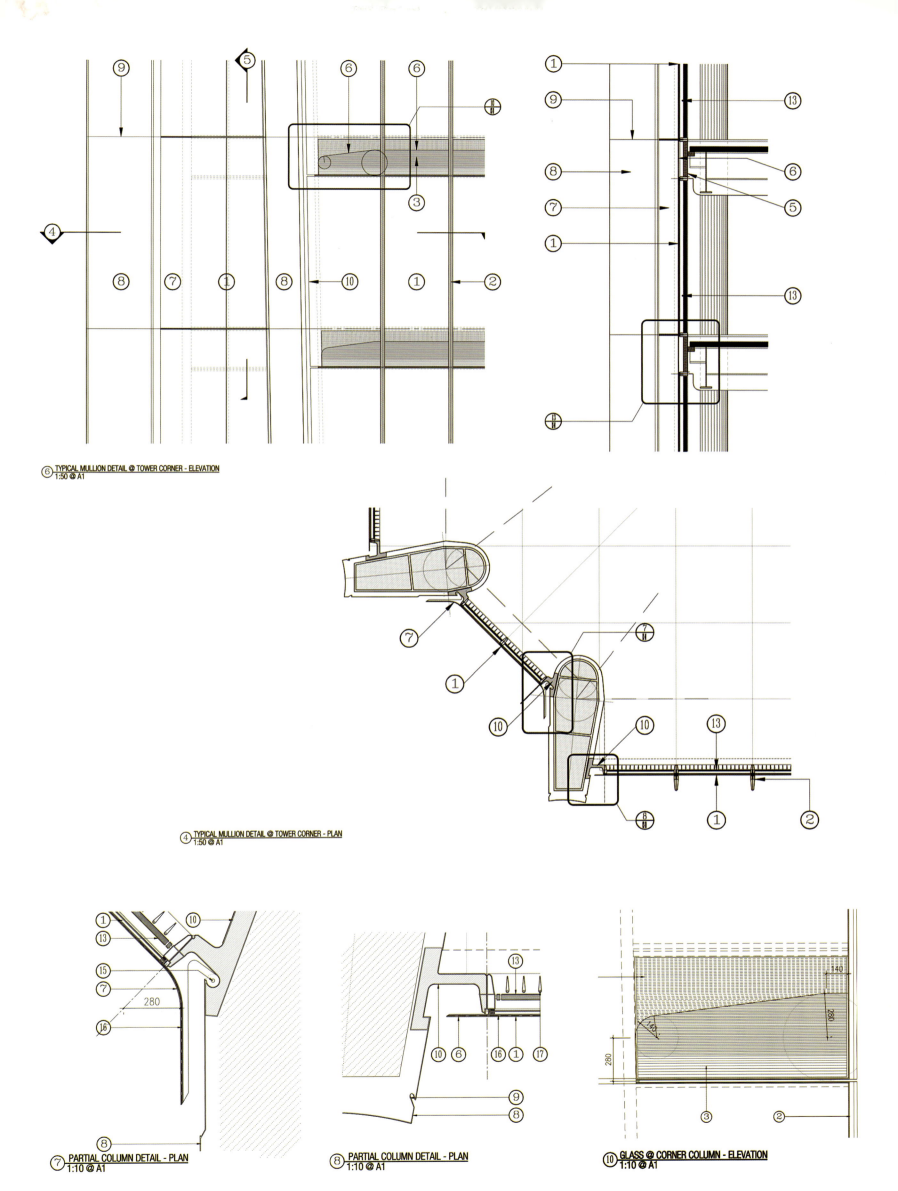

⑥ TYPICAL MULLION DETAIL @ TOWER CORNER - ELEVATION
1:50 @ A1

④ TYPICAL MULLION DETAIL @ TOWER CORNER - PLAN
1:50 @ A1

⑦ PARTIAL COLUMN DETAIL - PLAN
1:10 @ A1

⑧ PARTIAL COLUMN DETAIL - PLAN
1:10 @ A1

⑩ GLASS @ CORNER COLUMN - ELEVATION
1:10 @ A1

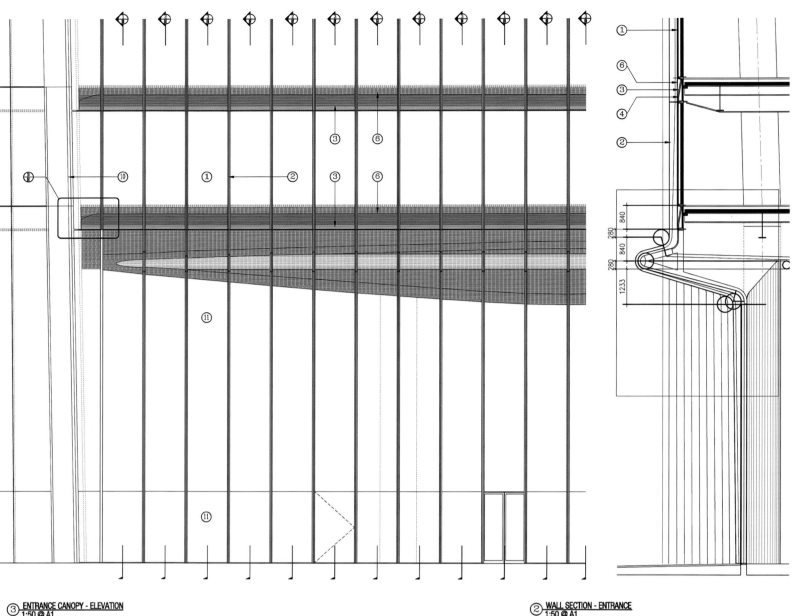

③ ENTRANCE CANOPY - ELEVATION
  1:50 @ A1

② WALL SECTION - ENTRANCE
  1:50 @ A1

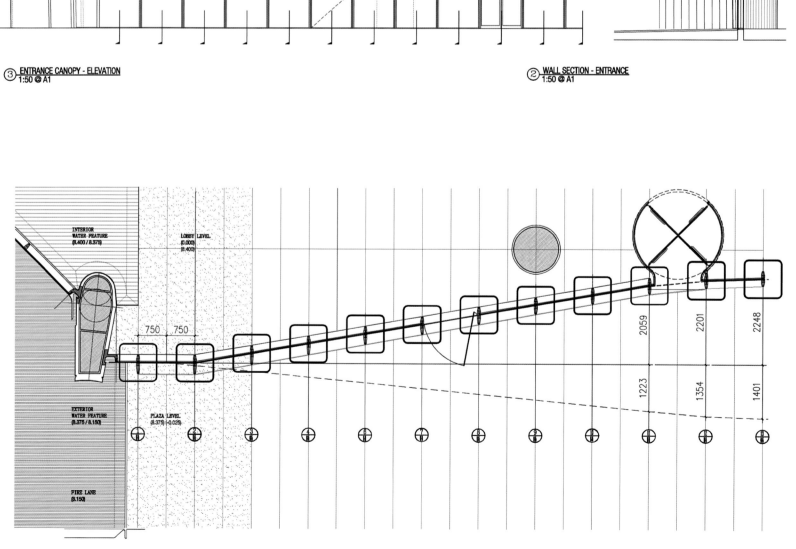

INTERIOR
WATER FEATURE
(8.400 / 8.375)

LOBBY LEVEL
(0.000)
(8.400)

EXTERIOR
WATER FEATURE
(8.375 / 8.150)

PLAZA LEVEL
(8.375) (-0.025)

FIRE LANE
(8.150)

750 750

1223 1354 1401

2059 2201 2248

① ENTRANCE CANOPY - PLAN
  1:50 @ A1

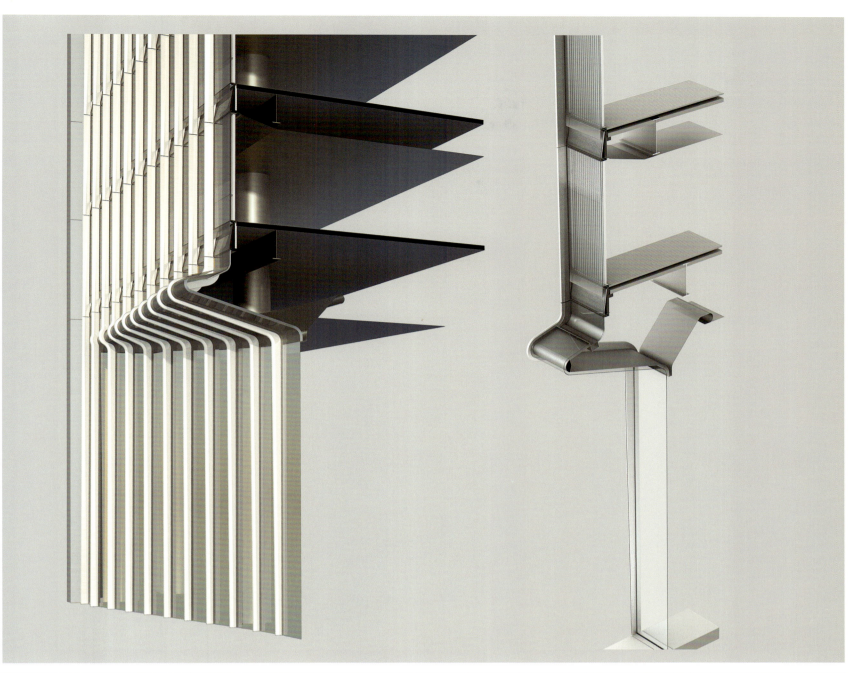

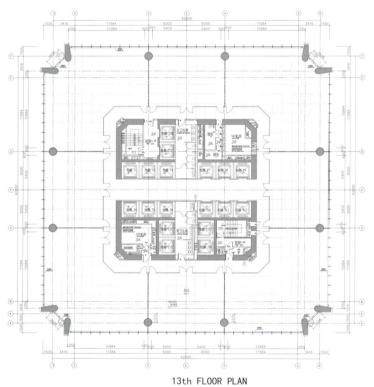

13th FLOOR PLAN

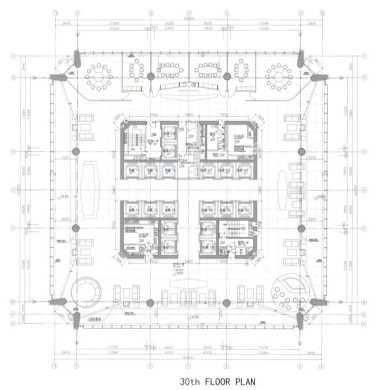

30th FLOOR PLAN

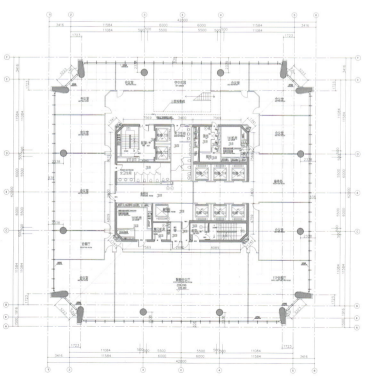

50th FLOOR PLAN

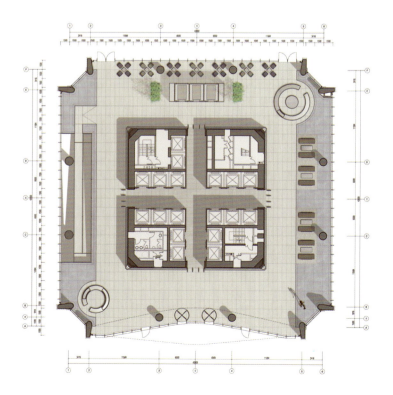

THE ONE

LOCATION:
HONG KONG,
CHINA

ARCHITECT: TANGE ASSOCIATES
ASSOCIATE FIRM: LWK & PARTNERS (HK) LIMITED
CONTRACTOR: GAMMON CONSTRUCTION LIMITED

CLIENT: CHINESE ESTATES GROUP
AREA: 40,000.00 m²
PHOTOGRAPHER: FREEMAN WONG

A 29-story mixed-use commercial tower incorporating retail, restaurants and cinemas, The One, is located on Nathan Road, one of the main streets in Hong Kong.

In the process of designing, Tange Associates always have Globalization and Localization, two opposing thoughts, in mind. Although the economic growth drives the internationalization of the city, they place importance on how they integrate a sense of place or country with architecture.

The site is located in the city of Hong Kong, where has been the center of finance and trading in Asia, while well-known city of shopping and gourmet in the world. They thought a sense of Hong Kong is various characters and spaces which people can experience with one's five senses, that is exactly what people can feel in Hong Kong, where lots of food-related shops such as the retail shops or food stalls stand side by side, comparing with large shopping malls. The facade's eye-catching layout is a vertical expression of Hong Kong's lively and energetic street life, applying a unique facade design to the different functions contained within the building.

The facade offers a different aspect of the building at night when the vibrant composition of the facade is animated by the illumination of the interior lights. As there has been the plan to make extensions to the hotel adjoining the site, we have received no request from the client for the facade on the border. However, out of consideration for people on the street and people living in the surrounding area, we proposed the simple, but muted and shiny champagne-tiled facade, instead of the cold huge concrete wall. The One, the tallest retail complex in Hong Kong and well recognized not only from the surrounding area but also from Hong Kong Island, is a new landmark on Nathan Road.

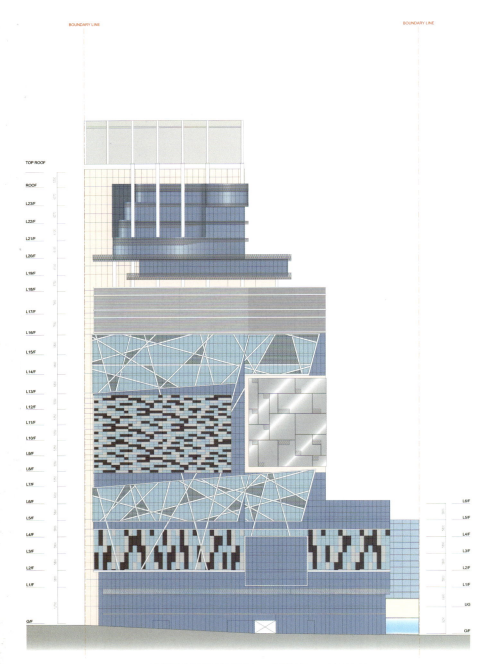

SOUTH ELEVATION   SCALE 1: 600(A3)

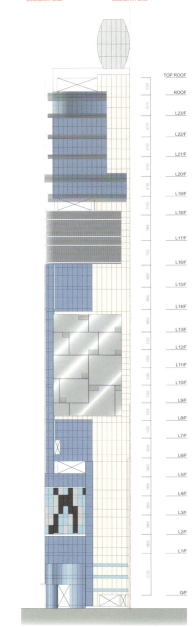

EAST ELEVATION

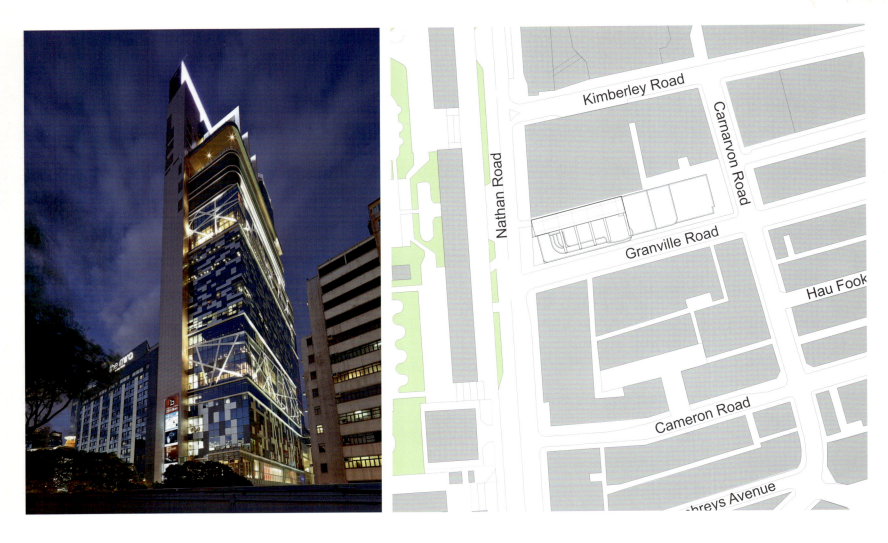

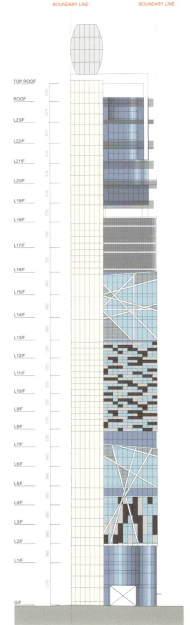

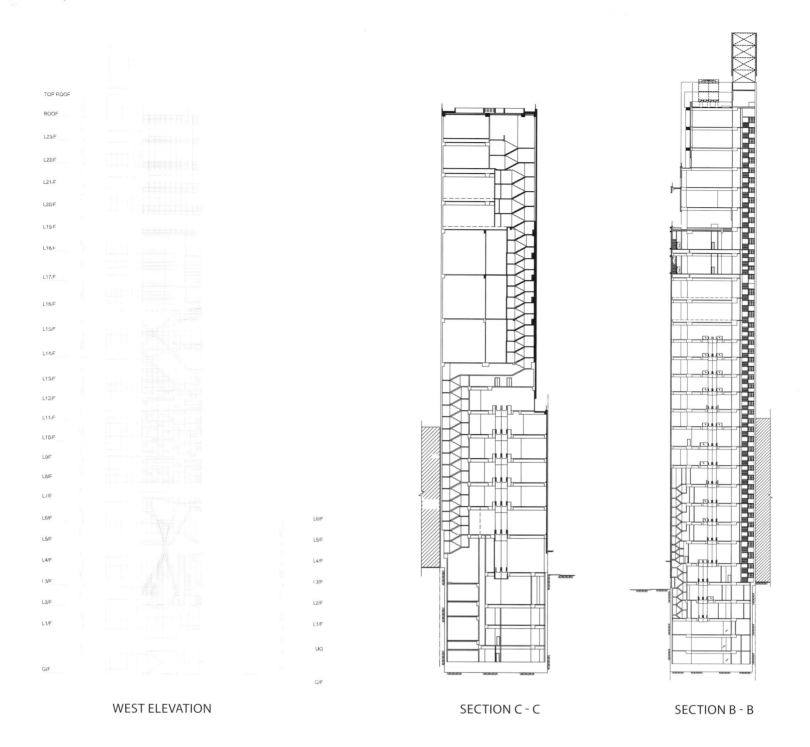

WEST ELEVATION

TOP ROOF
ROOF
L23/F
L22/F
L21/F
L20/F
L19/F
L18/F
L17/F
L16/F
L15/F
L14/F
L13/F
L12/F
L11/F
L10/F
L9/F
L8/F
L7/F
L6/F
L5/F
L4/F
L3/F
L2/F
L1/F
G/F

SECTION C - C

L6/F
L5/F
L4/F
L3/F
L2/F
L1/F
UG
G/F

SECTION B - B

The One 楼高 29 层，是一座混合用途的商业大楼，包括商场、餐厅和影院三部分。大厦位于弥敦道——香港的主要街道之一。

在设计过程中，丹下都市建筑设计公司已经有全球化和本地化两种不同的想法。虽然经济增长带动了城市的国际化，但设计师他们更注重在建筑中整合一种地方或国家的感觉。

大厦位于香港，这里一直都是亚洲的金融和贸易中心，并且在世界上以购物和美食而闻名。他们认为香港是一个拥有各种各样人物和空间的地方，人们可以用他们的五官体验香港，而这正是香港给人的真正感觉。在这里，大量与食品相关的商铺，如零售店或大排档并排存在，这是其与大型购物中心的不同之处。大厦醒目的外观布局是香港充满生机和活力的街头生活的直接表达，独特的外观设计满足了建筑物内不同功能的需要。

晚上，内部明亮的灯光使得外立面的组成部分变得生动活泼，给人不同的视觉。设计早已计划好了要扩建大厦毗邻的旅馆，但却迟迟没有收到客户对该外立面的具体要求。然而，出于对街道上行人和周边地区居民的考虑，设计师提议使用简单、柔和而有光泽的香槟酒色瓷砖外墙，而不是寒冷巨大的混凝土墙。The One，作为香港最高的零售商场，是周边地区和香港岛公认的弥敦道上的一个新地标。

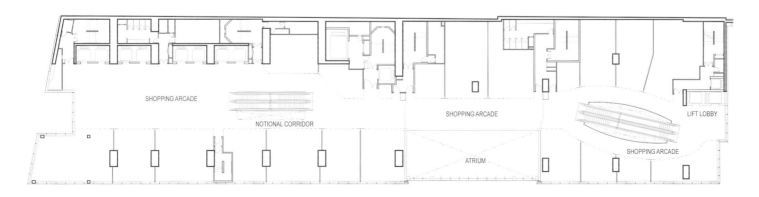

UPPER GROUND 2 (UG2) FLOOR PLAN

LOWER GROUND 2 (LG2) FLOOR PLAN

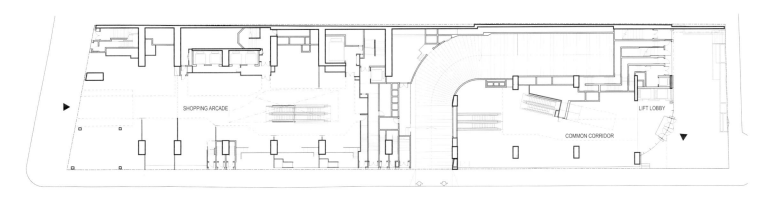

GROUND FLOOR PLAN

LEVEL 2 (L2) FLOOR PLAN

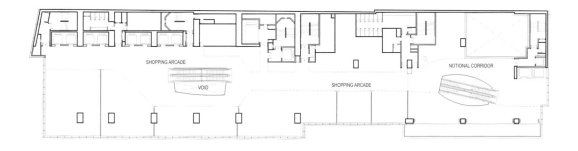

LEVEL 4 (L4) FLOOR PLAN

SCALE : 1/500 (A3)

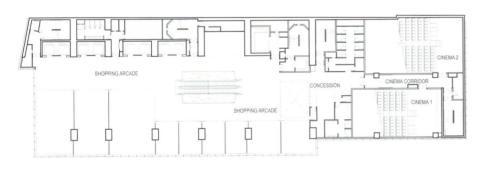

LEVEL 8 (L8) FLOOR PLAN

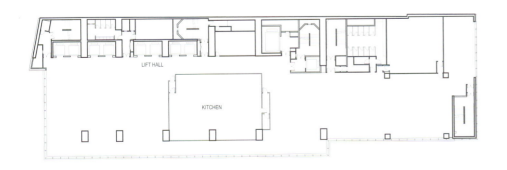

LEVEL 13 (L13) FLOOR PLAN (RESTAURANT)

LEVEL 18 (L18 RESTAURANT) FLOOR PLAN

MG TOWER

MG 大厦

LOCATION:
GHENT, BELGIUM   **ARCHITECT:** JASPERS-EYERS ARCHITECTS   **CLIENT:** MG REAL ESTATE

The shell of the entire tower was constructed with prefab elements around a central concrete pillar, with a facade of glass and 4 cm thick Chinese granite blocks.

For the design of the garden concrete slabs were installed on different levels in order to create a natural relief with a difference in height of 6 meters.

For the MG Tower to receive a building permit from the city of Ghent the plan also needed to include a clear vision on how traffic problems would be avoided. In that respect the MG Tower only has a pedestrian and delivery entrance on the high street.

All cars are deducted via the ring road into the underground parking garage. Suppliers will be allowed to access the landing quays via that same high street.

The underground parking lot has 430 parking spaces. A bicycle and pedestrian bridge leads to the especially for the tower installed new bus stop and further on to the terminus of the tram at neighbour IKEA. This hopefully will encourage employees to use public transport.

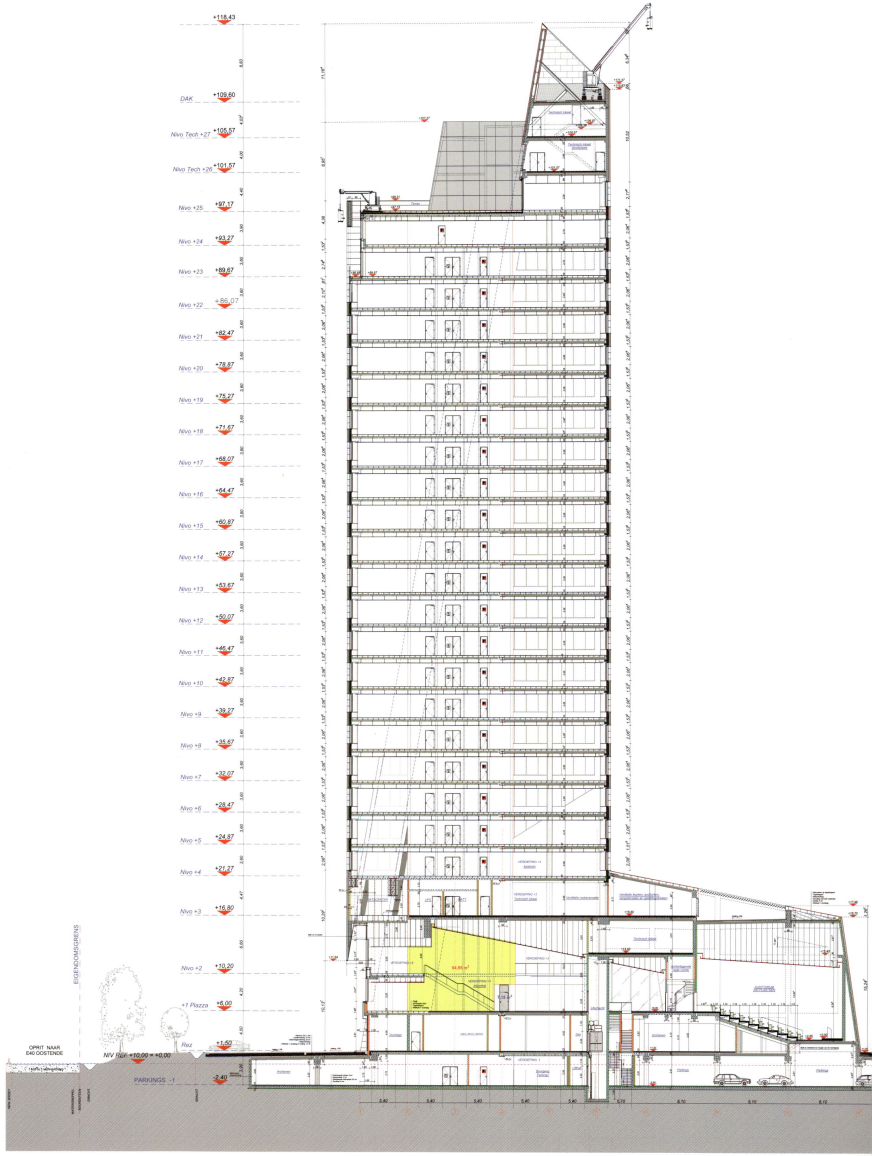

整座大楼外壳由预制件构成，中间是混凝土支柱，外立面为玻璃和 4 厘米厚的花岗岩。

为了设计出花园的效果，每一层都安装了混凝土板，创造出了 6 米高的天然浮雕。

为了获得建筑许可，MG 大厦的设计必须清楚地考虑到如何避免引起交通问题。基于这样的考虑，MG 大厦只在一条街上留出了行人通道和运输通道。所有的车辆都通过一条环形车道，直接进入地下停车库。供应商可使用在街上预留的运输通道。

地下停车库有 430 个停车位。一座过街天桥可以让自行车和行人从大厦到达附近的公交车站和附近 IKEA 的电车总站，这样的设计会鼓励员工使用公共交通工具。

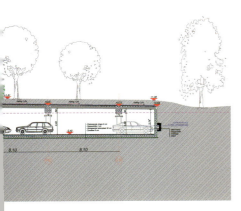

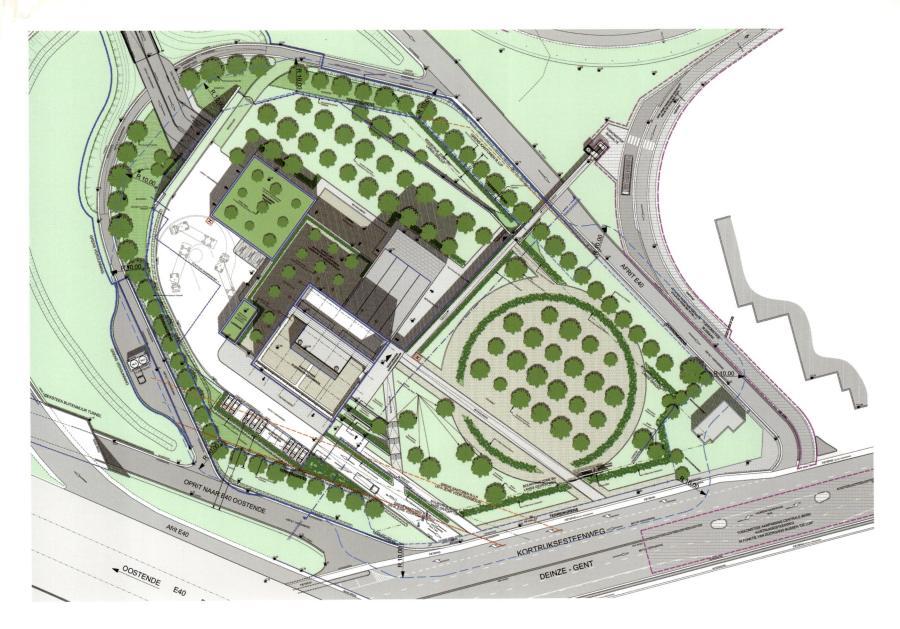

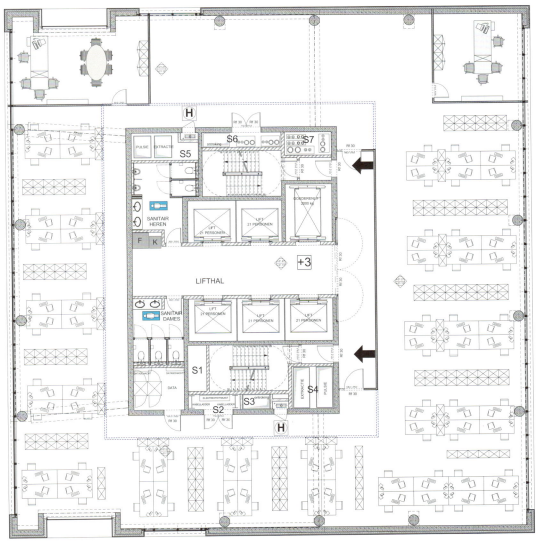

MAAS TOWER

马斯塔

LOCATION:
ROTTERDAM,
THE NETHERLANDS

ARCHITECT: DAM & PARTNERS ARCHITECTEN

PROJECT TEAM: DIEDERIK DAM, CEES DAM, CHRISTIAAN KOOP, MISCHA VAN EEKELEN, MARC STOOP, MATHIEU VAN EK, MILKO VAN MEEL, VALENTIJN OOSTBURG

IN COLLABORATION WITH: ODILE DECQ BENOIT CORNETTE ARCHITECTS, OTH ARCHITECTEN

AREA: 69,000 m²

PHOTOGRAPHER: LUUK KRAMER, MATHIEU VAN EK, ROB HOEKSTRA

With a height of 165 meters, the Maas Tower situated on the south bank of the Maas River where the Erasmus bridge touches land, forms the apex of the Kop van Zuid district. The base of the building is a anthracite-colored basalt plinth that rises from the water like a pier. The facades above this basalt base feature an aluminium skin, with the high tower's color changing from charcoal to silvery white as it ascends. The monochrome lower segment is in keeping with the adjacent row of buildings. A glass crystal-shaped volume that accommodates reception spaces lies on the waterfront at the base of the building. The central entrance on Laan op Zuid has a prominent folded glass porch. An ascending ramp through the center of the lobby lends access to a ten-storey public car park on the second floor. From the thirteenth storey on, the actual office building begins with 17 floors in both towers. An additional 15 floors continue in the highest section, that ends in a 7-meter high board room on the 45th floor, with a panoramic view of the Maas River.

ELEVATION SOUTH

ELEVATION EAST

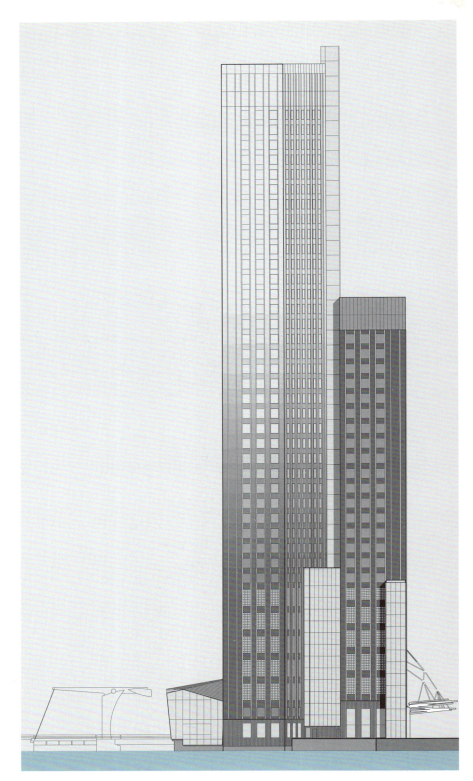

ELEVATION NORTH

马斯塔高 165 米，位于鹿特丹马斯河的南岸，紧邻伊拉斯谟斯桥，是鹿特丹 Kop van Zuid 区最高的建筑。大楼的煤灰色玄武岩底座驻扎在水中，像一个码头。大楼的外观幕墙采用铝材，从底座向上延伸，颜色逐渐由煤灰色过渡到银白色。建筑下半段的单色调与周围的建筑浑然一体。底座靠近河边的位置有一个玻璃水晶状的空间，该空间被用作接待区。位于 Laan op Zuid 的正门入口处设有推拉式玻璃门廊。中央大厅的中部有一个向上延伸的斜坡，可以到达第二层的公共停车场，停车场总共 10 层。从第 13 层开始，才是真正的办公区，两栋大楼的办公区都是 17 层。再往上还有 15 个楼层，第 45 层的董事室挑高 7 米，可以看到马斯河的全景。

ELEVATION WEST

SECTION A

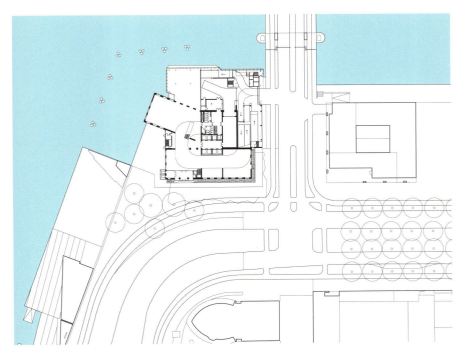

SITE PLAN

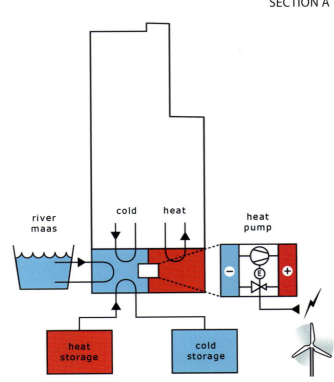

river
maas

cold   heat

heat
pump

heat
storage

cold
storage

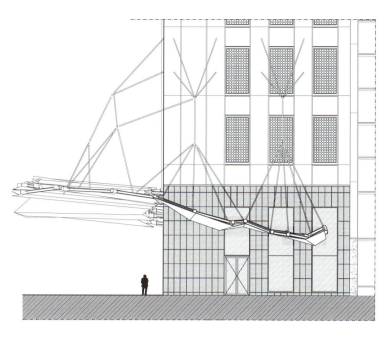

ELEVATION SOUTH_CANOPY

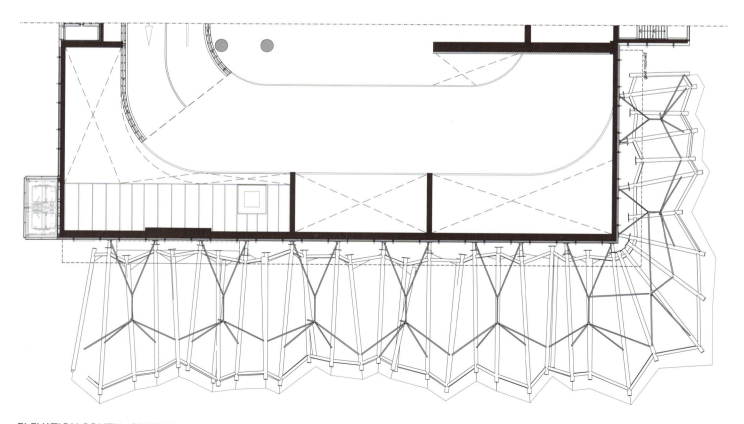

ELEVATION SOUTH_CANOPY

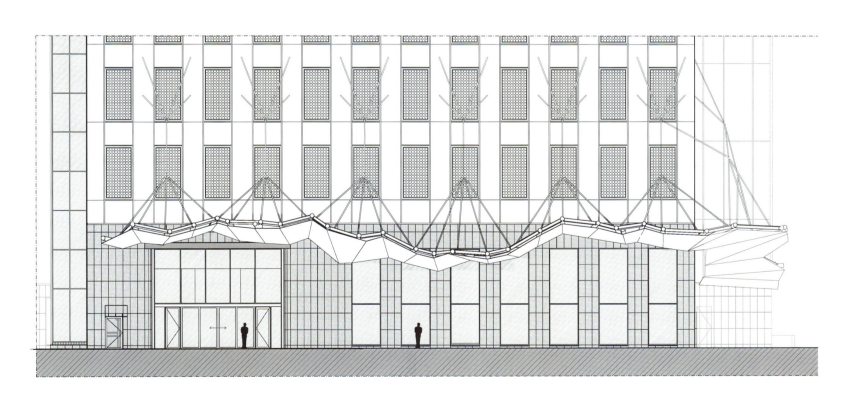

ELEVATION WEST_ENTRY WITH CANOPY

1 / 200   0   2   4m

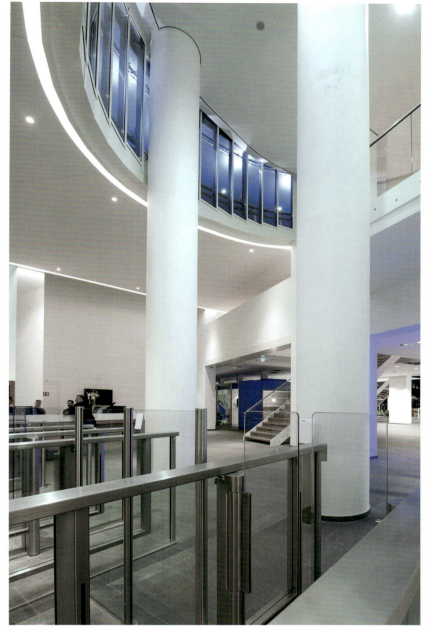

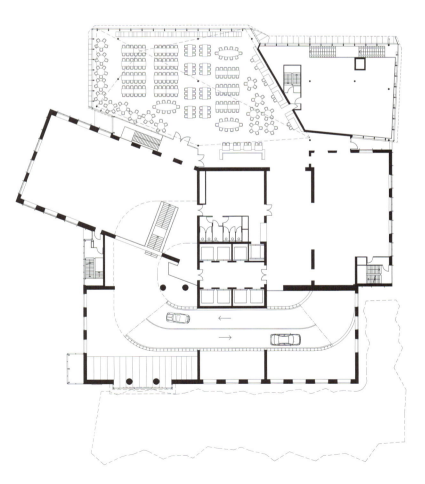

GROUND FLOOR PLAN

1 / 400

0   4   8m

FIRST FLOOR PLAN

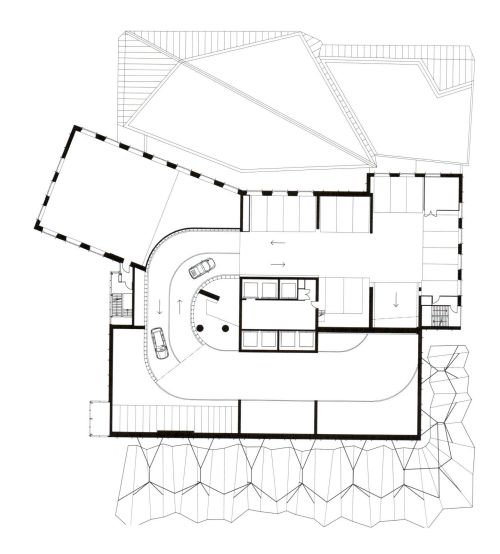

SECOND FLOOR PLAN

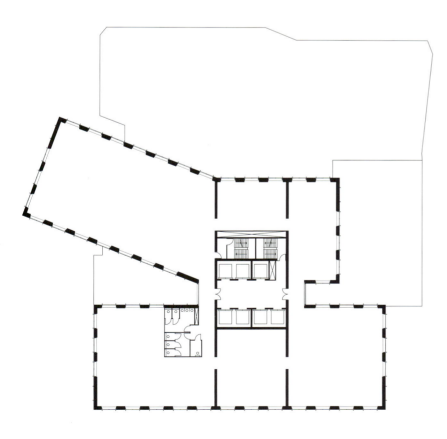

TYPICAL FLOOR PLAN LEVELS

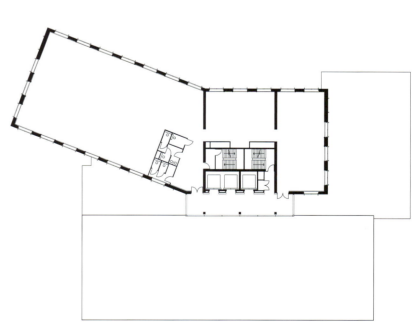

TYPICAL FLOOR PLAN HIGH-RISE_LEVEL 31~43

INDEX
索 引

TVSDISIGN

tvsdesign, a business corporation wholly owned by its employees, is an internationally recognized design firm that provides architecture, interior design and planning services. The firm has full service offices in Atlanta, Chicago, Dubai and Shanghai. They draw upon the knowledge and experience of many market segments including corporate/commercial office, higher education, retail, public assembly, hospitality and sustainability. Founded in 1968, the firm has steadily expanded its practice and currently employs 125+ professionals.

CHRISTOPH MÄCKLER

Christoph Mäckler is the head of the Prof. Christoph Mäckler Architekten architectural office in Frankfurt am Main and has been Professor for Urban Planning at the Technische Universität Dortmund since 1998. He founded the German Institute for Civic Art in 2008 and is also active as a consultant for many cities. His recently completed buildings and current projects include OpernTurm and Tower 185 in Frankfurt am Main, the Zoofenster high-rise development in Berlin, the Augustiner Museum in Freiburg im Breisgau and Terminal 3 at Frankfurt Airport.

GHD

GHD is one of the world's leading engineering, architecture and environmental consulting companies. The cornerstone of their business is their client-centred culture and teamwork based approach known as One GHD.
Wholly-owned by its people, GHD is dedicated to understanding and helping their clients achieve their goals in the global markets of water, energy and resources, environment, property and buildings, and transportation. Their network of 6500 professionals collaborate to improve the built, economic and social environment of the communities in which we operate.
GHD is committed to sustainable development, safety and innovation. They care for the wellbeing of our people, assist communities in need and conduct busines.

R&AS ARCHITECTURE

Carlos Rubio Carvajal and Enrique Alvarez-Sala are partners since 1980 .
Both together have obtained many awards in the world among many others. They are currently developing a skyscraper of 400 meters high in Saudi Arabia. They have also been selected for numerous national and international exhibitions, such as the Venice Architecture Biennale 2004 or the traveling exhibition "35 + Building Democracy: Social Spanish Architecture ". Their work has been published in the most prestigious national and international publications.

TANGE ASSOCIATES

At Tange Associates, the aim is to create architecture within an urban environment that is geared towards comfort. A place where people feel they want to live, work and visit, again and again. Their practice and projects span the globe, but large or small, regardless of borders, their designs focus on meeting the unique needs and comfort of their clients while adapting to the surrounding environment. To fully understand the clients' needs and aspirations they engage them in continuous, active dialogue. The result is unparalleled, creative and comfortable spaces – inspired by their clients.

FENTRESS ARCHITECTS

Fentress Architects is a global design firm that passionately pursues the creation of sustainable and iconic architecture. Together with their clients, Fentress creates inspired design to improve the human environment. Founded by Curtis Fentress in 1980, the firm has designed US$27 billion in architectural projects worldwide, visited by over 300 million people each year. Fentress is a dynamic learning organization, driven to grow its ability to design, innovate and exceed client expectations. The firm has been honored with more than 400 distinctions for design excellence and innovation, and in 2010, Curtis Fentress was recognized by the American Institute of Architects with the most prestigious award for public architecture, the Thomas Jefferson Award. Fentress has studios in Denver, Colorado; Los Angeles, California; San Jose, California; Washington, D.C.; London, U.K.; and Shanghai, China.

ALLESWIRDGUT

Since 1997, AllesWirdGut has been working on projects of different scales – from urban strategies to interior design. The pragmatic approach searches for the potential of the given context. By looking at so-called problems as chances, new unexpected possibilities come into existence. The goal beyond the given task is to find additional qualities in order to realize them. The four architects of AllesWirdGut met at the Technical University of Vienna (TU) where first common projects were created. What distinguishes the team is the input of diversified characters as well as the co-operation without hierarchies and without specialisations. The actual formula for success of the young architects is team spirit. The building task is analized thoroughly, nothing is forbidden on principle, a lot is possible. The logic answer at first sight is not always the best. As a result surprising methods of resolutions occur for both the builder and the architects. Those methods are reviewed and perfected mutualy.

CARLOS ZAPATA STUDIO

Carlos Zapata Studio is an award winning master planning, architectural design, interior design and furniture design firm whose projects range from professional sports arenas to airport concourses to commercial high rises to private homes. Working on projects around the globe, CZS specializes in finding functional solutions to difficult programmatic issues, while maintaining a commitment to elegance and design excellence.

The work of Carlos Zapata Studio has won numerous design awards and been featured in books and magazines around the world. Carlos himself has received several prestigious awards, including being named by US News and World Report as "Someone to Watch" at the beginning of his career; as well as being selected as one of Interior Design magazine's "30 Under 30" as well as their "40 Under 40".

DAM & PARTNERS ARCHITECTEN

Dam & Partners Architecten, founded in 1962 and led by Cees Dam and Diederik Dam, occupies itself with a wide range of projects differing in both nature and scale, from complex urban developments, offices and shopping centers, theatres and town halls, hotels, restaurants and leisure facilities, restoration of historical buildings, social housing and luxurious villas to interiors, furniture, carpets and glass.

Dam & Partners Architecten has developed its own unique architectural philosophy and style that can be moulded to the concept and context of each commission. This results in a style that can undergo refreshing changes in appearance under the influence of the latest insights into sustainability, of new technologies, new opinions about design and the design process, and the intensive involvement of the firm and its employees in the arts and the cultural debate. Throughout this ongoing process, however, architecture in all its craftsmanship and with all its tradition and laws of dimensions, scale and rhythm remains the primary focus.

MING LAI ARCHITECTS INC

MING LAI ARCHITECTS INC provides service to large-scale mix-use development projects, office buildings, star hotels, service apartments, corporation headquarters and many more. Its distinctive style hints of the design philosophy of master architect Mies Van der Rohe. To the premise of rationality and economics, it is able to incorporate local culture and traditions with the use of technological advances and ideas of sustainable construction to create simple, but elegant architecture.

DIRK LOHAN

Dirk Lohan, together with Floyd D. Anderson, is a founder of Lohan Anderson, and is responsible for the firm's designs and long-term development.

The firm's philosophy is that successful architecture must respond not only to economic constraints but also to larger social and physical conditions. Looking to the future, Mr. Lohan's philosophy is becoming ever more relevant due to the rapid changes in technology, the growing importance of environmental concerns and the impact of globalization in the life and work of all people.

Recent projects include a 45-story office tower at 353 N Clark Street in Chicago, a large field office complex for the FBI Chicago and a headquarters building for Calamos Financial Services in Naperville, Illinois.

Throughout his career, Mr. Lohan has successfully humanized the traditional modernism of the fifties and sixties by infusing variety and texture to enrich building design. Dirk Lohan's achievements in architecture have been recognized with many awards including being appointed to Fellow in 1983 by the American Institute of Architects.

ATKINS

Atkins is one of the world's leading design consultancies. They have the breadth and depth of expertise to respond to the most technically challenging and time-critical projects and to facilitate the urgent transition to a low carbon economy. Their vision is to be the world's best design consultant.

Whether it's the architectural concept for a new supertall tower, the upgrade of a rail network, master planning a new city or the improvement of a management process, they plan, design and enable solutions.

With 75 years of history, 17,700 employees and over 300 offices worldwide, Atkins is the world's 13th largest global design firm (ENR 2011), the largest global architecture firm, the largest multidisciplinary consultancy in Europe and UK's largest engineering consultancy for the last 14 years. Atkins is listed on the London Stock Exchange and is a constituent of the FTSE 250 Index.

LEIGH & ORANGE

Founded in 1874, Leigh & Orange is a large, well-established architectural company offering architectural, structural and civil engineering services to major local and overseas clients in both public and private sectors. In more recent years, L&O has grown into an established international practice concentrating on the business of architecture and interior design. Headquartered in Hong Kong, L&O has offices in Beijing, Shanghai, Fuzhou and the Middle East with over 400 staff.

L&O certified under ISO 9001 in respect of its management system in 1996 and has since obtained certification for its environmental approach to ISO 14001 and for Health & Safety to OHSAS 18001.

VERITAS DESIGN GROUPAS

VERITAS Design Group is a multi-disciplinary consulting firm which provides integrated architecture, planning, interior design, landscape design, quantity surveying and environment consulting services for innovative buildings and spaces throughout the Asian region and beyond.

VERITAS was founded in 1987 in Kuala Lumpur, Malaysia, upon the principles of constant innovation and a commitment to service quality. During its early years, VERITAS was the fastest growing design practice in its market, emerging as one of the region's leading firms both in terms of size of practice and scale of projects. VERITAS is one of South-East Asia's foremost design firms. Leading the firm is Group President David Mizan Hashim, who is supported by Group Principals Lillian Tay, Azif Nasaruddin, Ng Yiek Seng, Azril Jaafar and Zainal Aziz. They are complemented by business unit Principals and backed by a team of over a hundred qualified professionals and over a hundred other support staff.

ROJKIND ARQUITECTOS

Rojkind Arquitectos is a Mexico City-based innovative firm, founded in 2002, whose work spans across a diverse and global platform providing solutions to contemporary architectural and urban strategies. They are nurtured by a multinational team, gathering experts from all fields that are involved in a highly collaborative, research and experimental based design process.

They are committed to achieving design excellence; but above all, they pursue a fascination for thorough investigation and exploration of contemporary programs, latest technologies and the translation of complex digital design into simple solutions for local fabrication. The result: a wide spectrum of design initiatives, from the intimacies of small objects to the complexities of large buildings and master plans.

By understanding that their profession is at constant flux (economical, political, social) they are willing to take risks and pursue uncommon paths by providing urban strategies that are able to go beyond what is expected or what is normally a given.

EMBA(ESTUDI MASSIP-BOSCH ARQUITECTES)

EMBA(ESTUDI MASSIP-BOSCH ARQUITECTES) is an internationally oriented architecture studio established by Enric Massip-Bosch in 1990 to provide comprehensive and highly personalized architecture design and urban planning. Their work follows a multi-disciplinary approach and integration of experiences and knowledge. Their work is developed in different countries and the core of the practice includes architects of different nationalities. This core incorporates external specialized collaborators depending on the specific needs of each project, maintaining permanent relations with their practice, both professionally and academically.

DBI DESIGN

DBI Design was incorporated in Australia in 1980. The core expertise of the firm was drawn from specialist hotel and resort firms based in California and Hawaii. During the development boom of that decade DBI Design quickly established a reputation for innovative design of large scale resorts and mixed use master plans, hotels and multi-unit residential projects. In addition to its impressive track record for the design of hotels, resorts and urban master planning, DBI is now a leader in tall (+100m) and super tall (+300m) building design and construction and is a recognized leader in sustainable high rise technology. DBI's collaborative and rigorous approach ensures the best possible outcomes for the client, community and environment. All of their projects seek to place the human being at the center of the built environment and deliver environmentally, culturally, socially and economically sustainable outcomes for our future.

WINGÅRDHS

Wingårdhs is today among the five largest architect groups in Sweden, and among the ten largest in the Nordic Region. This falls well in line with the target set in the office's general objectives that they must be one of Scandinavia's leading architecture firms. Gert Wingårdh, architect SAR/MSA and CEO, is the owner, manager and responsible architect in all projects.

The office has been operating in Gothenburg since 1977, in Stockholm since 1985, and in Malmö since 2011. The office works on all types of projects, from product development and interior design to large structures and urban planning.

STUDIO DANIEL LIBESKIND

Daniel Libeskind established his architectural studio in Berlin, Germany in 1989 after winning the competition to build the Jewish Museum in Berlin. In February 2003, Studio Daniel Libeskind moved its headquarters from Berlin to New York City when Daniel Libeskind was selected as the master planner for the World Trade Center redevelopment. Daniel Libeskind's practice is involved in designing and realizing a diverse array of urban, cultural and commercial projects internationally. The studio has completed buildings that range from museums and concert halls to convention centers, university buildings, hotels, shopping centers and residential towers. In addition to the New York headquarters, Studio Libeskind has European partner offices based in Zurich, Switzerland and Milan, Italy.

INGENHOVEN ARCHITECTS

Christoph Ingenhoven founded the architectural studio ingenhoven architects in 1985. The studio is located in Duesseldorf's "Media Harbour" and employs some 90 architects, interior architects, designers, draughts men and model makers from nineteen different countries.

The practice is well known for its open work atmosphere that emphasizes teamwork and communication. Essential for ingenhoven architects's work are an ecological and sustainable approach, the wellbeing of the users, technical innovation, flexibility and efficiency, logical structures and a precise finish. With offices in Switzerland, Australia, Singapore and the USA the high quality of execution can be guaranteed.

SAMOO ARCHITECTS & ENGINEERS

SAMOO Architects & Engineers is an architectural design firm headquartered in Seoul, Korea with diversified services including architectural design, urban planning, interior design, engineering, and construction management services. Since the firm's founding in 1976, SAMOO has developed into one of the world's largest architectural firms with a number of branch offices worldwide. Employing more than 1,000 professionals, SAMOO is involved in a diverse portfolio including office, civic, cultural, healthcare and biotechnology, residential, hospitality, academic, high-tech industrial, transportation, super-tall buildings and mixed-use projects. SAMOO was awarded the "red dot Design Award" in 2011 and ranked 9th worldwide by BD (UK) World Architecture TOP 100 and 2nd among architectural design firms of the Pacific Rim in 2012.

JAEGER AND PARTNER ARCHITECTS LTD

Jaeger and Partner Architects Ltd. is a unique design practice comprised of talented architects and planners whose broad experience and international backgrounds allow them to provide comprehensive design services. With principal offices in China, Italy and Korea, they have realized projects internationally of various scale and complexity. Services range from small scale interventions to urban planning, from architecture to interiors, and have been recognized for the highest quality of design.

Excellent design integrates the timeless principles of appropriateness, proportion, attention to craft and detail, with progressive solutions holistically embracing tomorrow in a truly sustainable approach. It stands for the synthesis of practical efficiency and aesthetic values, giving design the capacity to endure. Jaeger and Partner is driven by the pursuit of design excellence in every project, believing that the quality of our surroundings has a direct effect on the quality of our lives.

SHIMIZU CORPORATION

Shimizu's more than 200-year history began in 1804, when founder Kisuke Shimizu launched a carpentry business in the Kanda Kajicho district in Edo (now Tokyo). During its early years, Shimizu appointed Eiichi Shibusawa, a renowned industrialist of the Meiji Era (1868~1912), as senior advisor and based its management style on his "Rongo to Soroban" ("The Analects and the Abacus"). This work set forth the concept of how businesses can contribute to society based on a firm union between ethics and economics, thereby achieving fair returns and attractive growth. Construction technologies and methods have advanced rapidly since that time. Still, the approach of striving to ensure growth based on quality products that satisfy customers remains the unchanging foundation for all Shimizu activities.

Each structure Shimizu builds incorporates ideas that are drawn from a wide range of parties while centering on the needs of their customers. Shimizu identifies these ideas and applies them in a way that breathes new life into its structures.

MEINHARD VON GERKAN

Meinhard von Gerkan, founding partner, born in 1935 in Riga/Latvia, Prof. Dr.h.c.mult., graduate architect, 1965 co-founder with Volkwin Marg of the architectural partnership of von Gerkan, Marg and Partners, 1974 appointed professor, chair A for Design at the Carolo-Wilhelmina Technical University in Brunswick, head of Institute A for Architectural Design, member of the Freie Akademie der Künste in Hamburg, 2005 honorary doctorate in design from Chung Yuan Christian University in Chung Li/Taiwan, 2007 honorary professor, East China Normal University College of Design, Shanghai/China, 2007 president of the Academy for Architectural Culture (aac). Numerous awards, including Fritz Schumacher Prize, Romanian State Prize, bronze plaque of the Freie Akademie der Künste, Hamburg, BDA Prize, Federal Cross of Merit.

NIKOLAUS GOETZE

Nikolaus Goetze, partner, born in 1958 in Kempen, graduate architect, partner at von Gerkan, Marg and Partners since 1998, management of the gmp offices in Hamburg Elbchaussee, Shanghai and Hanoi, projects in Asia include Shenzhen International Convention and Exhibition Center, International Convention and Exhibition Center, Nanning, Guangzhou Development Central Building, National Conference Center, Hanoi, Lingang New City, Dalian Twin Towers, Hanoi Museum, Siemens Center, Shanghai, National Assembly House, Hanoi.

后记

本书的编写离不开各位设计师和摄影师的帮助，正是有了他们专业而负责的工作态度，才有了本书的顺利出版。参与本书的编写人员有：Tvsdesign, Carlos Rubio Carvajal, Enrique Álvarez-Sala Walther, Rafael Vargas, Mark Bentley, Juan José Mateos Bermejo, Tange Associates, Koji Horiuchi, Carlos Zapata Studio, CHAPUIS Tristan, Murray Alan, GHD, Brendan Texeira, Alex Atienza, Atkins, Leigh & Orange Ltd, Shu He, DBI Design, Gert Wingårdh, Karolina Keyzer, Wingårdh Arkitektkontor AB, Åke E:son Lindman, Tord-Rickard Söderström, Ola Fogelström, Prof. Christoph Maeckler Architekten, Thomas Eicken, Christian Richters, von Gerkan, Marg und Partner, RTKL Associates Inc., David Whitcomb, Meinhard von Gerkan, gmp architekten, EMBA(ESTUDI MASSIP-BOSCH ARQUITECTES), Enric Massip-Bosch, Aleix Antillach, Elena Guim, Jon Ajanguiz, Esteve Solà, Ricardo Mauricio, Carlos Cachón, Cornelia Memm, Cristina Feijoo, Heidi Reichenbacher, Rita Pacheco, Rodrigo Vargas, Jana Alonso, Marta Marcet, Mariana Arámburu, Ca Immo, SAMOO Architects & Engineers, G.S Architects & Associates, Yum Seung Hoon, Ingenhoven Architects, Fentress Architects, Nick Merrick, Pawel Sulima, Dominique Perrault Architecture, Ming Lai Architects, Paul Noritaka Tange, Yasuhiro Ishino, Kazuya Ishida, Kenichi Matsuda, Philippe Iliffe-Moon, Kentaro Shiratani, Marc Tey, AllesWirdGu, Lohan Anderson LLC, Bill Zbaren, Lillian Tay, Jacky Lee, Iskandar Razak, Lau Kok San, Paul Gadd, Studio Daniel Libeskind, rojkind arquitectos, Jaspers-Eyers Architects, Shimizu Corporation, Dam & Partners Architecten, Luuk Kramer, Mathieu van Ek, Rob Hoekstra, Jaeger and Partners Architects, Renzo Piano Building Workshop

DOMINIQUE PERRAULT

Figure of French architecture, Dominique Perrault gained international recognition after having won the competition for the National French library in 1989. This project marked the starting point of many other public and private commissions abroad, such as The Velodrome and Olympic swimming pool of Berlin, the extension of the European Court of Justice in Luxembourg, the Olympic tennis center in Madrid, the campus of Ewha's University in Seoul and the Fukoku Tower in Osaka. On-going projects include works as the DC towers in Vienna, the rehabilitation of the former mechanical engineering halls and the central library as well as the construction of the Teaching Bridge of the Ecole Polytechnique Fédérale in Lausanne, and the renovations of the Pavillon Dufour at the Chateau de Versailles and of the Longchamp Racecourse in Paris.

MG REAL ESTATE

MG Real Estate is a real estate development and Investment Company specialized in the development of large logistic schemes, warehouses, retail, residences and state-of-the-art offices in prime locations. MG Real Estate is a Belgian privately held group of companies with a strong equity basis and a lean organizational structure, which enables them to take swift decisions.

Together with the in-house development and construction teams, MG Real estate will customize client's project according to their own specific requirements; following an approach that respects client's budget and which ensures that the project will be completed within the established deadline.

All developments exceed the competitors' standards on quality and architectural design.

Each of their developments is supported by the in-depth knowledge of the international real estate investors' standards. This results in a Total Cost of Ownership approach and the focus on a land bank in prime locations only. In that aspect MG Real Estate is currently also active in the Netherlands, France, Luxembourg and Germany.

RTKL

A worldwide architecture, engineering, planning and creative services organization. Part of the ARCADIS global network since 2007, RTKL specializes in providing its multi-disciplinary services across the full development cycle to create places of distinction and designs of lasting value. RTKL works with commercial, workplace, public and healthcare clients on projects around the globe.

CCDI

Founded in 1994, CCDI is a large architectural consulting firm that provides integrated services for urban construction and development. They have business units covering broad industry sectors to achieve specialization. Five regional offices that are located in Shanghai, Beijing, Shenzhen, Chengdu and New York, as well as branches and representative offices in Chongqing, Nanjing, Tianjin, Wuhan, Xi'an, and Kunming, set up their operations cross-regionally. Successful cases of their company include National Aquatics Center – Water Cube, National Tennis Center, Shenzhen Ping'an International Financial Center, Rockbund Shanghai, Jinan Olympic Sports Center, Shenzhen Dameisha Vanke Center, Hangzhou Olympic Sports Center, Harbin West Railway Station, Harbin Taiping International Airport, Kunlun Apartment CBD Beijing, City Crossing Shenzhen Phase II, Tianjin Vanke Crystal City and Tianjin International Cruise Home Port. Since its founding, CCDI has made remarkable technological advancements in independent innovation and environmental protection.

THE RENZO PIANO BUILDING WORKSHOP

The Renzo Piano Building Workshop (RPBW) was established in 1981 by Renzo Piano with offices in Genoa, Italy and Paris, France. The Workshop has since expanded and now also operates from New York.

The Workshop permanently employs about 100 architects together with a further 30 support staff including CAD operators, model makers, archivers, administrative and secretarial staff.

They provide full architectural design services and consultancy services during the construction phase. Their design skills also include interior design services, town planning and urban design services, landscape design services and exhibition design services.

The firm is licensed to practice architecture in France. Renzo Piano personally is registered as an architect in France and Italy.

The Workshop has successfully undertaken and completed projects around the world. Exhibitions of Renzo Piano and the Building Workshop's works have been held in many cities worldwide, including a major retrospective exhibition in 2007 at the Milan Triennale.

ACKNOWLEDGEMENTS

We would like to thank everyone involved in the production of this book, especially all the artists, designers, architects and photographers for their kind permission to publish their works. We are also very grateful to many other people whose names do not appear on the credits but who provided assistance and support. We highly appreciate the contribution of images, ideas, and concepts and thank them for allowing their creativity to be shared with readers around the world.